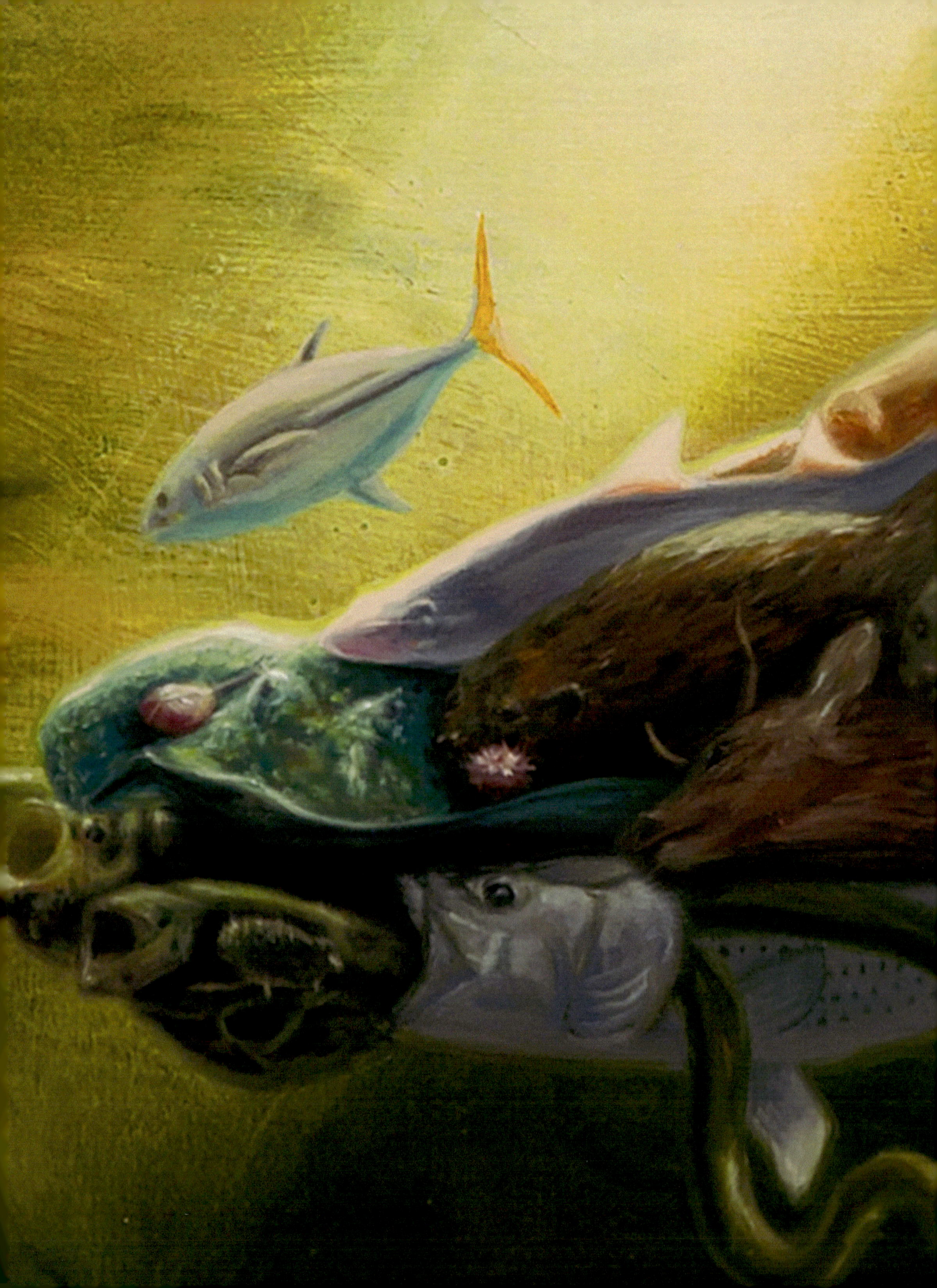

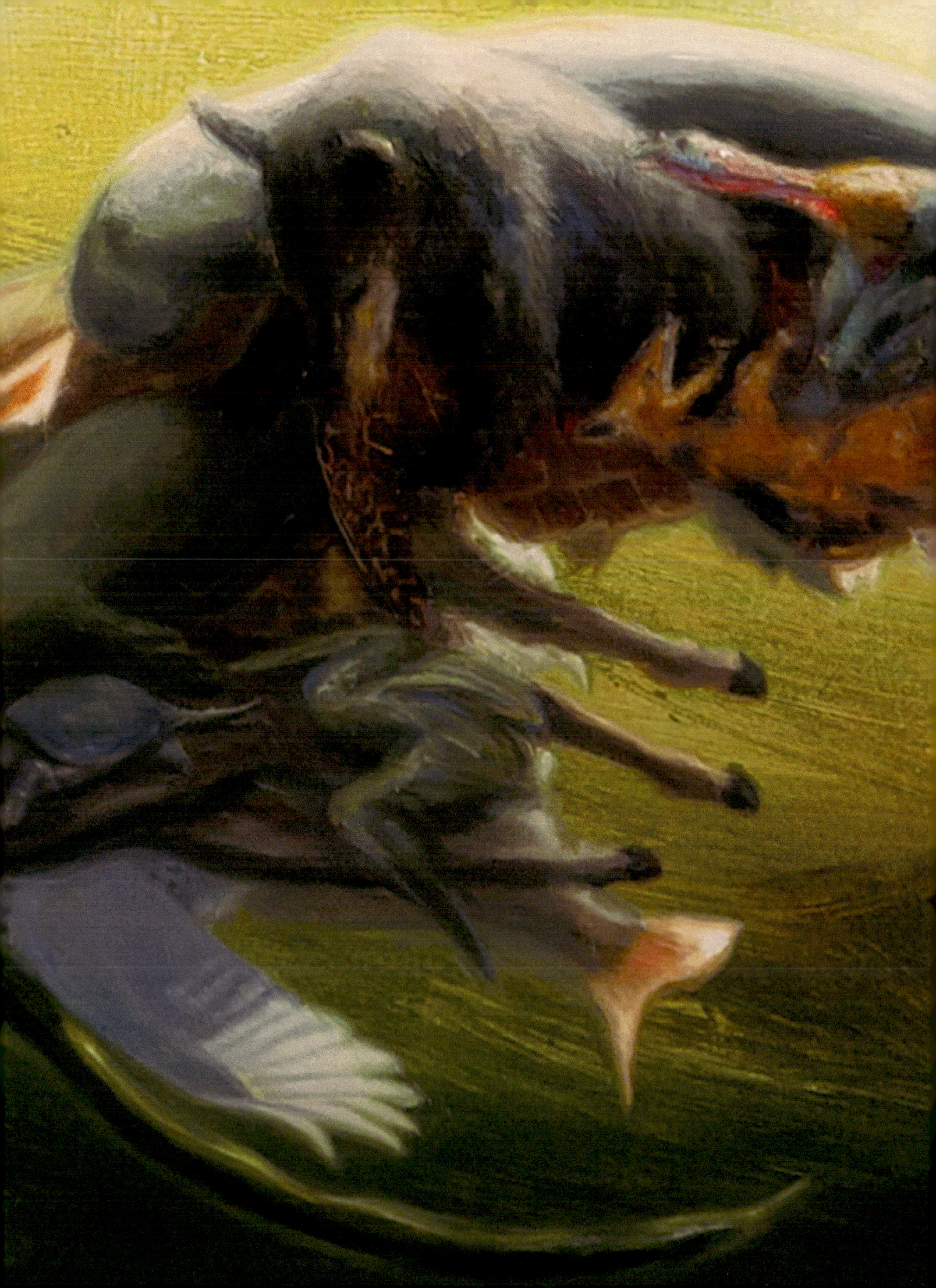

ECO-URGENCY

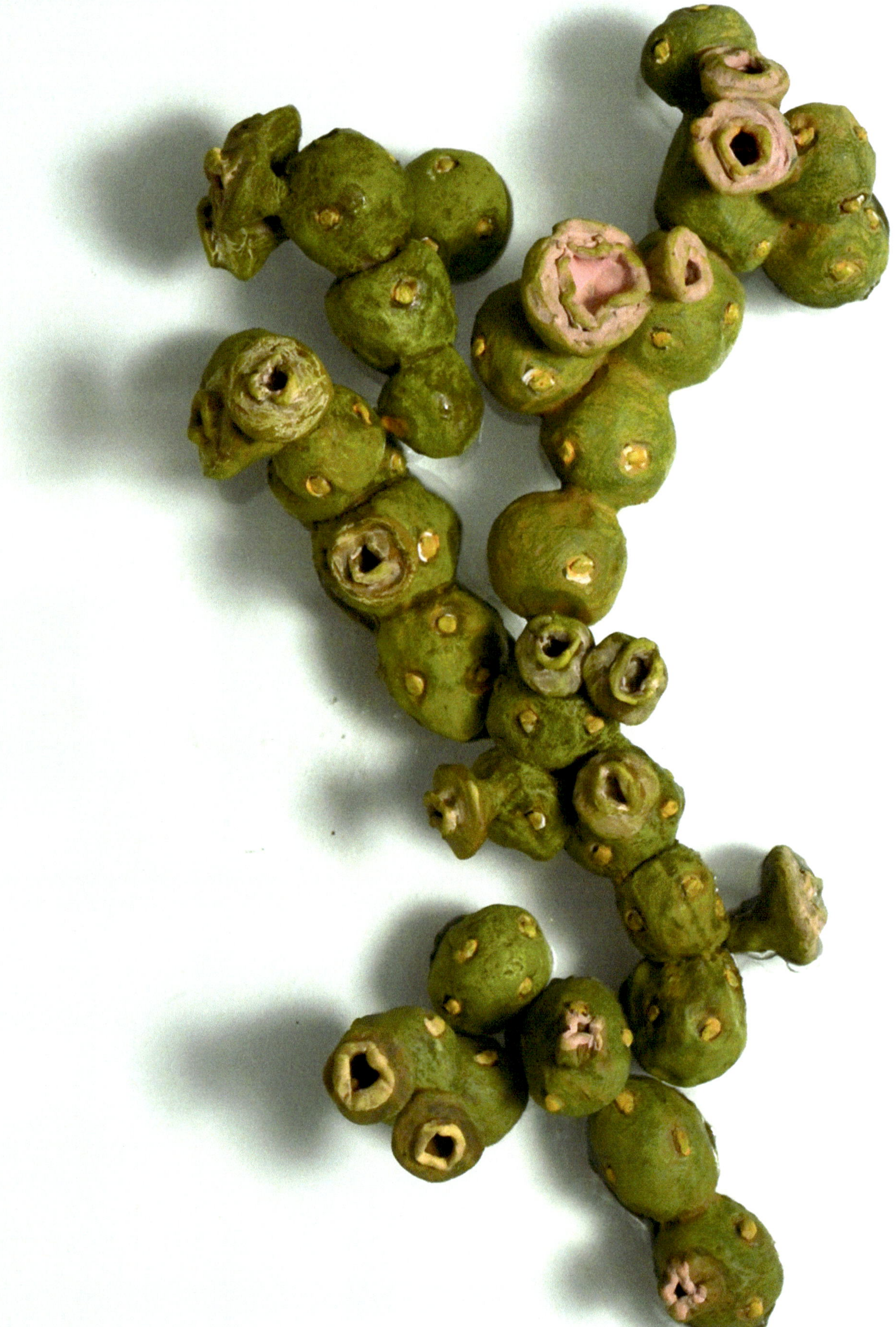

ECO-URGENCY
NOW OR NEVER

Lehman College Art Gallery   City University of New York

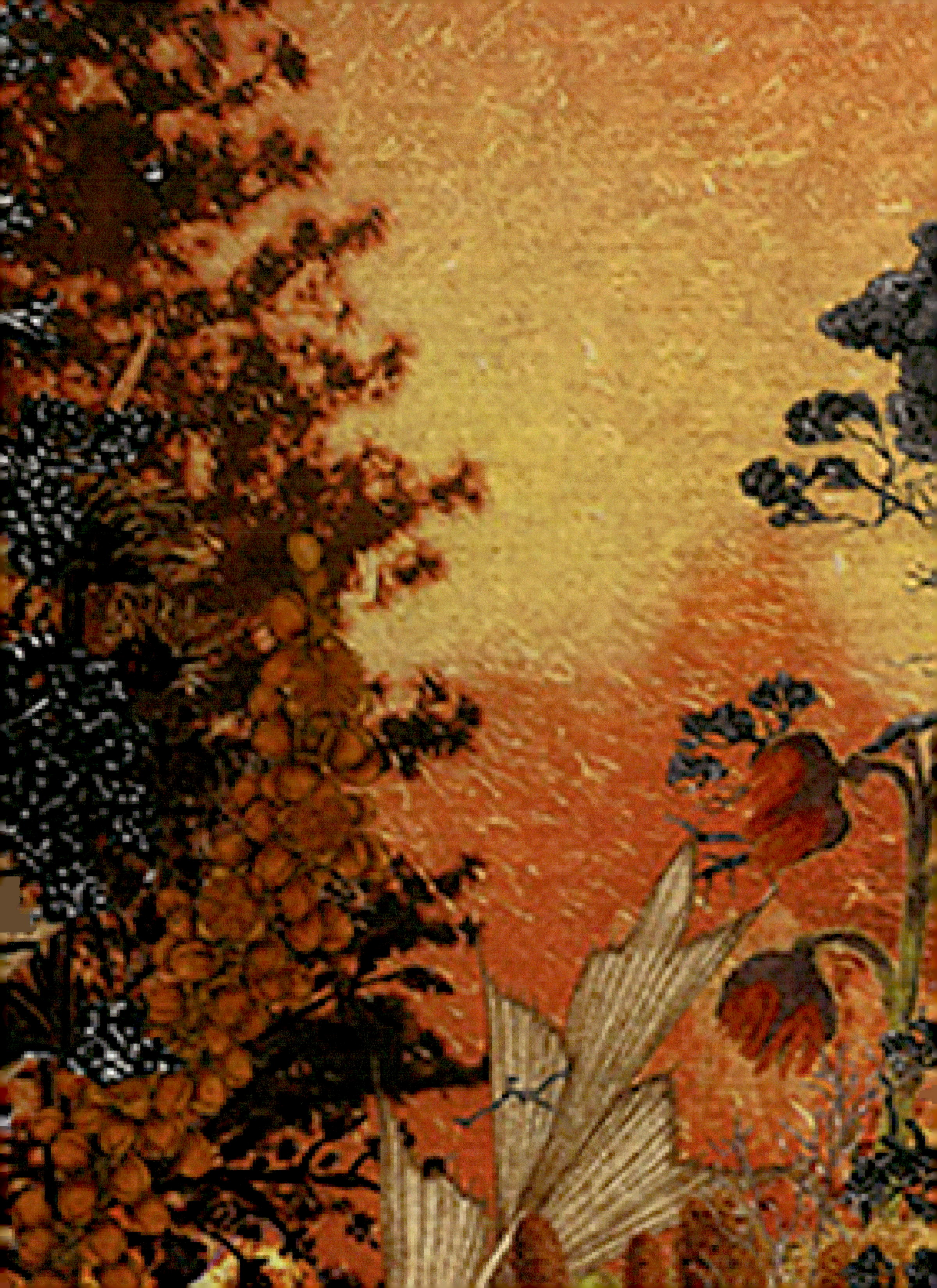

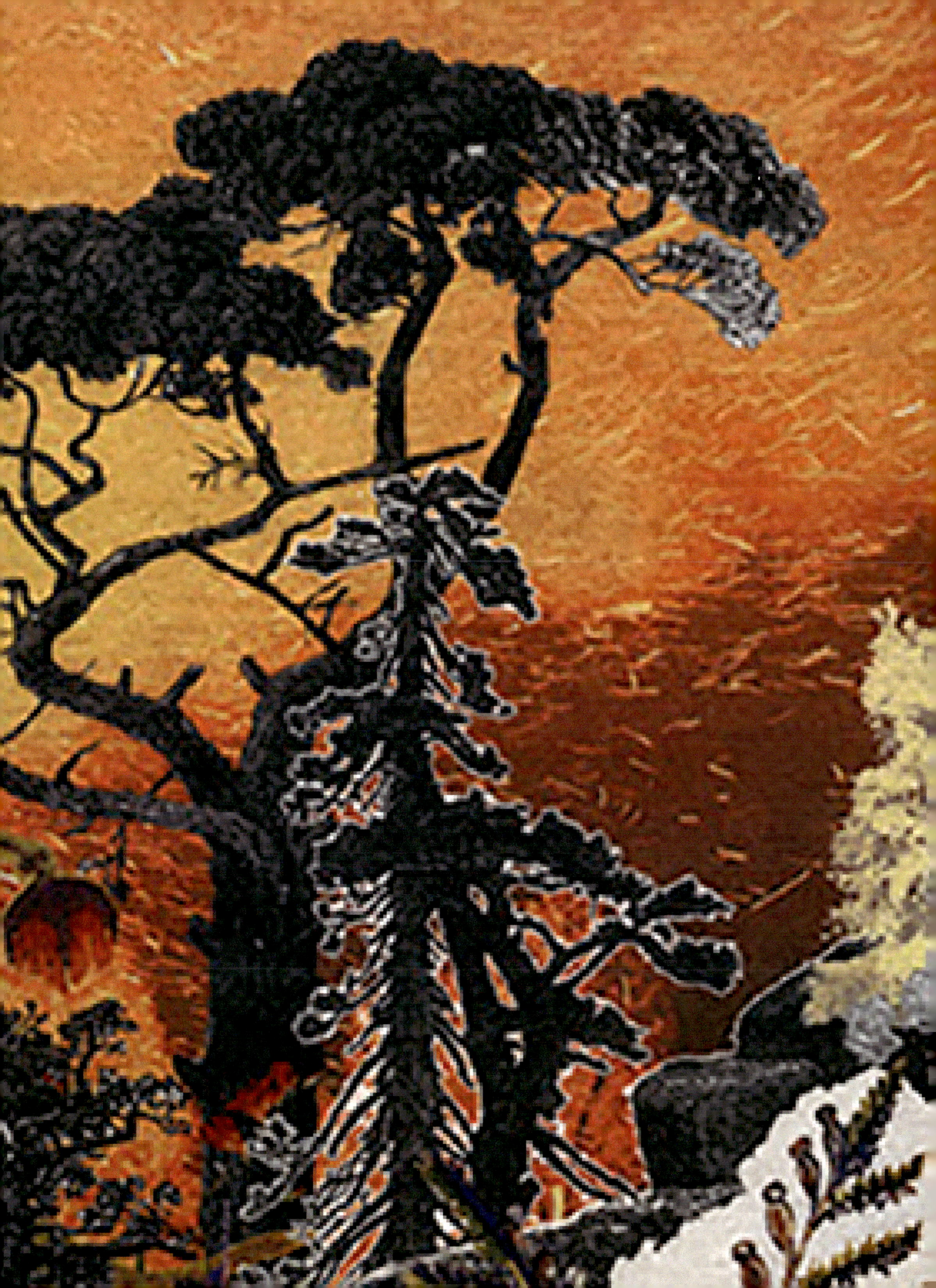

This catalog is published on the occasion of the online exhibition *Eco-Urgency Now or Never* presented by the Lehman College Art Gallery/ The City University of New York, December 7, 2021 – April 23, 2022.
Wave Hill Glyndor Gallery, Part I August 28 – December 5, 2021.
Lehman College Art Gallery, Part 2 December 4, 2021 – April 23, 2022.

ISBN: 978-1-7343095-6-0

LEHMAN COLLEGE ART GALLERY
250 Bedford Park Boulevard West
Bronx, New York 10468
www.lehmangallery.org

Design Michelle Frank

Copy Editor Linda Locke

Installation Photography Cary Whittier

Cover and Inside Covers: Alexis Rockman,
GOWANUS, 2013 (Detail)

Title Page: Rachel Frank,
THRESHOLDS, CHOLLA CHAIN FRUIT CACTUS I, 2018

Following Title Page: Francesco Simeti,
REFUGIUM (CYPRUS FIRE), 2019 (Detail)

Copyright Page: Rachel Frank,
THRESHOLDS, 2019 (Video Still, Detail)

Artists Page: Samantha Box,
UNTITLED 3, 2021 (Detail)

Introduction Page: Alicia Grullon,
PERCENT FOR GREEN BILL PROTEST SIGN, 2014 (Detail)

Catalog of the Exhibition Page: Will Wilson,
AUTO IMMUNE RESPONSE (AIR) No.5, 2004 (Detail)

Back Page: Samantha Box, **PORTABLE HOMELAND #6,** 2021 (Detail)

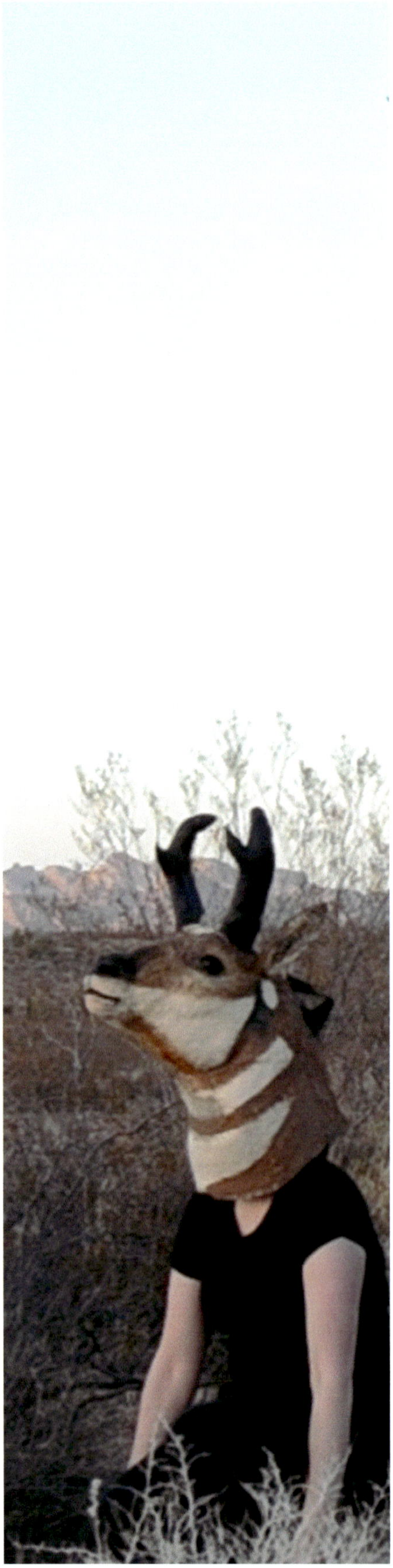

ACKNOWLEDGMENTS

EXHIBITION SUPPORTERS

New York City Department of Cultural Affairs in partnership with the City Council

New York City Council through the Honorable Andrew Cohen and the Bronx Delegation

New York State Council on the Arts with the support of Governor Andrew Cuomo and the New York State Legislature

The Robert Lehman Foundation

The Edith and Herbert Lehman Foundation

The Jarvis and Constance Doctorow Family Foundation

The Milton and Sally Avery Arts Foundation

The New Yankee Stadium Community Benefits Fund

Catalog Entries: Bartholomew F. Bland, Deborah Yasinsky
Researcher: Ashley Lum

Eco-Urgency: Now or Never is co-organized by Glyndor Gallery, Wave Hill Public Garden and Cultural Center, Bronx, New York and Lehman College Art Gallery, The City University of New York, and is curated by Bartholomew Bland, Gabriel de Guzman, Jesse Bandler Firestone, Eileen Jeng Lynch and Deborah Yasinsky. The curatorial concept for the exhibition was initiated with Jennifer McGregor, former Director of Arts at Wave Hill.

The Lehman College Art Gallery exhibition program is supported by the NYC Department of Cultural Affairs in partnership with the New York City Council; the New York City Council and the Bronx Delegation; the Edith and Herbert Lehman Foundation; The Jarvis and Constance Doctorow Family Foundation; and The New Yankee Stadium Community Benefits Fund.

Support for Wave Hill's Visual Arts Program is provided by Con Edison; the Council for Canadian American Relations; Horace W. Goldsmith Foundation; Milton and Sally Avery Arts Foundation; the National Endowment for the Arts; New York City Department of Cultural Affairs; New York Community Trust Edward and Sally Van Lier Fund; New York State Council on the Arts with the support of Governor Kathy Hochul and the New York State Legislature; The Pollock-Krasner Foundation; Stavros Niarchos Foundation; Teiger Foundation; The Willem de Kooning Foundation; and by the Cathy and Stephen Weinroth Commissioning Fund for the Arts. The City of New York owns the buildings and grounds of Wave Hill. Wave Hill's operations are made possible in part by public funds provided through the New York City Department of Cultural Affairs; the Zoos, Botanical Gardens and Aquariums Grant Program administered by the New York State Office of Parks, Recreation and Historic Preservation; the National Endowment for the Arts; the New York State Department of Environmental Conservation; and New York State Council on the Arts with the support of Governor Kathy Hochul and the New York State Legislature.

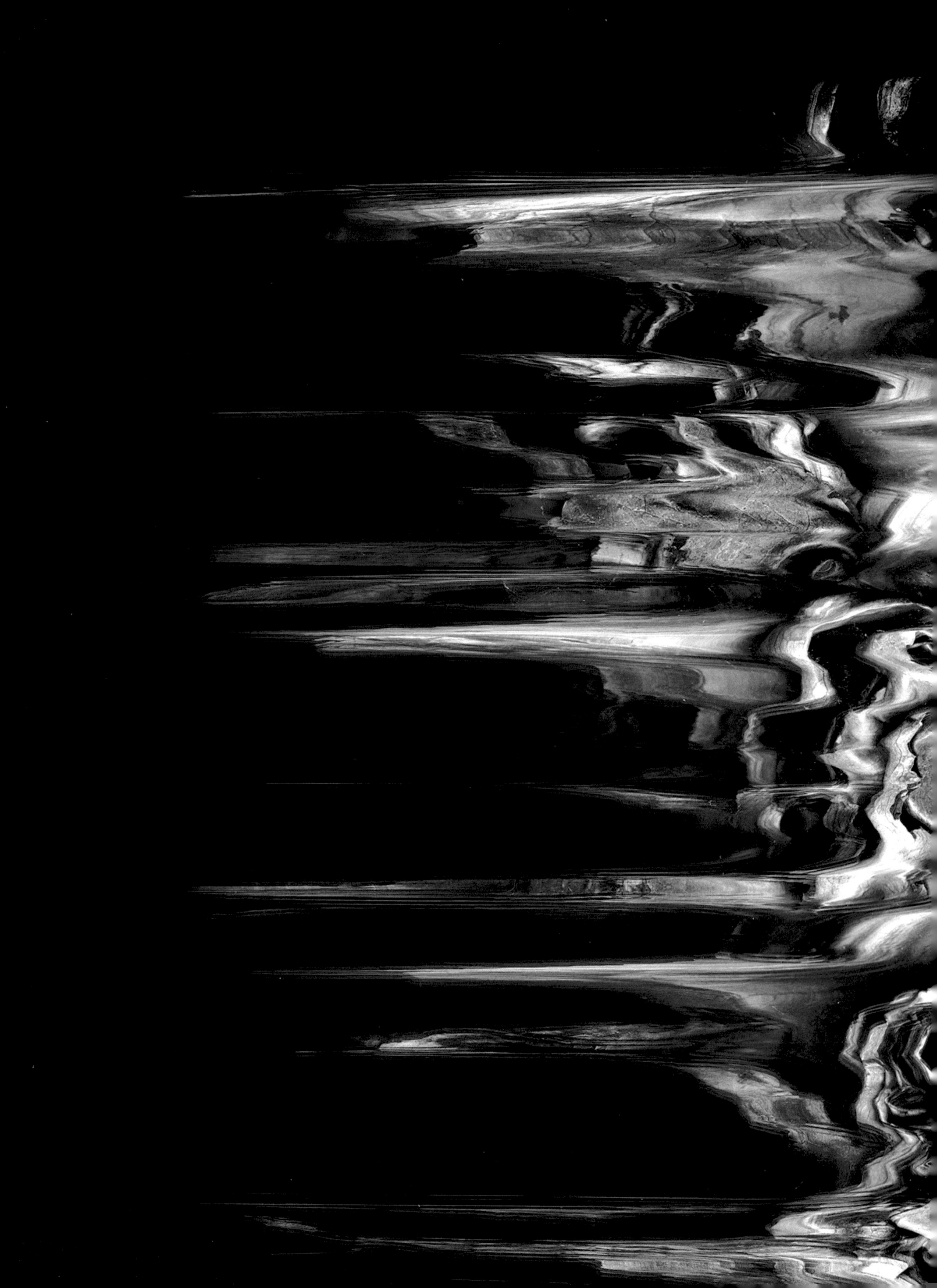

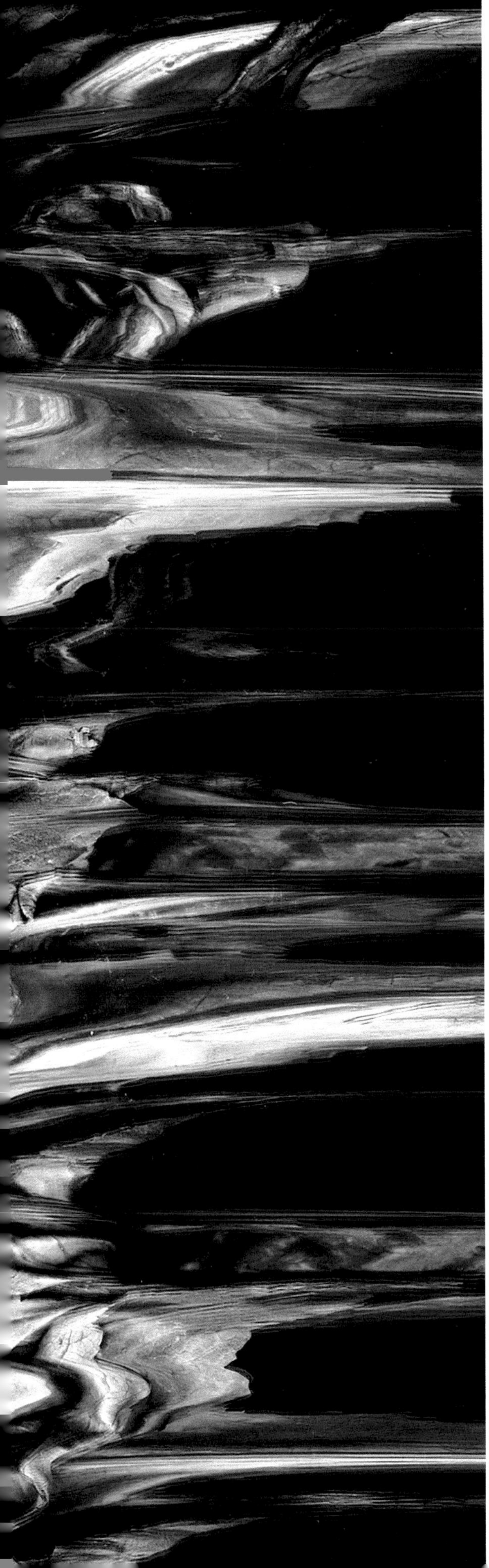

ARTISTS

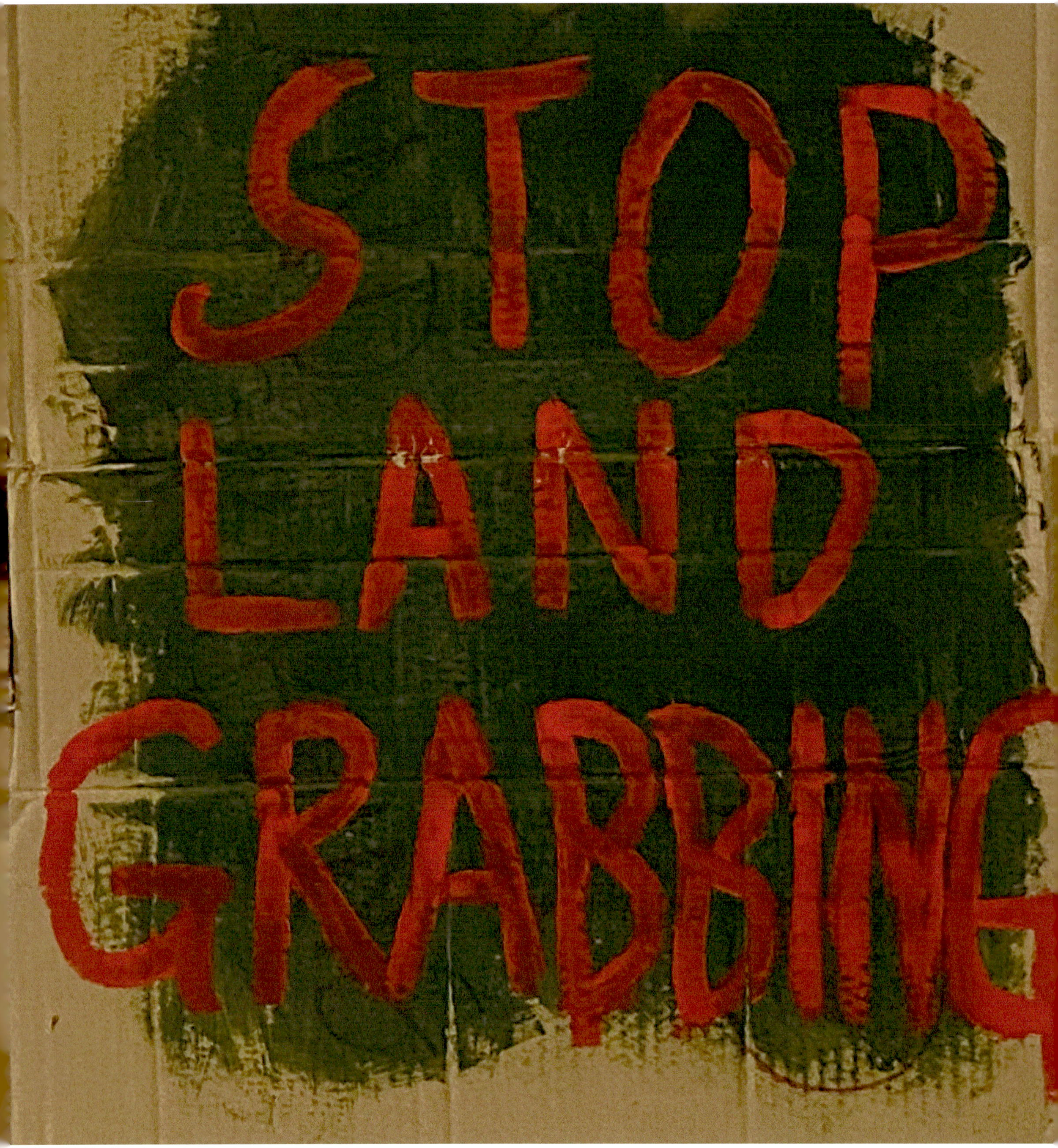
STOP
LAND
GRABBING

INTRODUCTION

Nature is all around us, making up the key elements of our surrounding environment, even in the densest neighborhoods of New York City. It surrounds the City's concrete roads, growing in pocket parks. It is alive in window boxes perched on the limestone sills of its older buildings and and in the grassy strips that lie between its modern glass skyscrapers.

As we go about our urban lives, how often do we think of the resources we consume? We are exposed to a steady, ominous drumbeat of news that tells us how the crisis of global warming already upon us will bring even greater problems in the decades to come. Many scientists argue that we are now living through the "Age of the Anthropocene," the geological period during which human activity and production have become the dominant influences in our environment, irrevocably altering what we think of as "nature." Yet on a crisp autumn day or a first day in spring that promises warmth and light, there is a cognitive disconnect: we know the planet is changing but it is difficult for us to grasp the enormity of the change.

Eco-Urgency: Now or Never brings together artists who examine the climate crises that looms larger each day. First presented, *Now*, at Wave Hill during Fall 2021 looked at the urgency of the present moment, raising our awareness of the climate and its immediate changes through a holistic approach that gave us an overview of its social, political, and environmental concerns. The subject of this volume, *Never* at Lehman, presented in Spring 2022, viewed ecological crises in endlessly repeating, mournful, cycles. These artists examine the resonant patterns of history that led us to this current moment of crisis, and look at possible futures.

INTRODUCTION

The exhibition investigates overlapping thematic inquiries that preoccupy contemporary artists, which include "Depicting Imminent Change" — the imagining of apocalyptic climate catastrophe; "Politics and the Land" exploring political and economic policies that affect natural resources; "Stewardship" focusing on the urge to collect, archive and preserve; "Decentering Human Experience" positing the importance of animal perspectives; and finally, "Grief," which presents artistic work that is elegiac and prompts questions about what it means to memorialize what is already lost or about to be lost. To encourage our engagement and deepen our awareness of these issues, the artists in *Eco-Urgency: Now or Never* rely on research, critical analysis, observation, and direct action that are manifest as objects, videos, performances, and community organizing. There are no easy answers, yet *Eco-Urgency: Now or Never* distills the current moment, so that it can be understood and acted upon, provoke conversation, and prompt us to share ideas.

Bartholomew F. Bland

Deborah Yasinsky

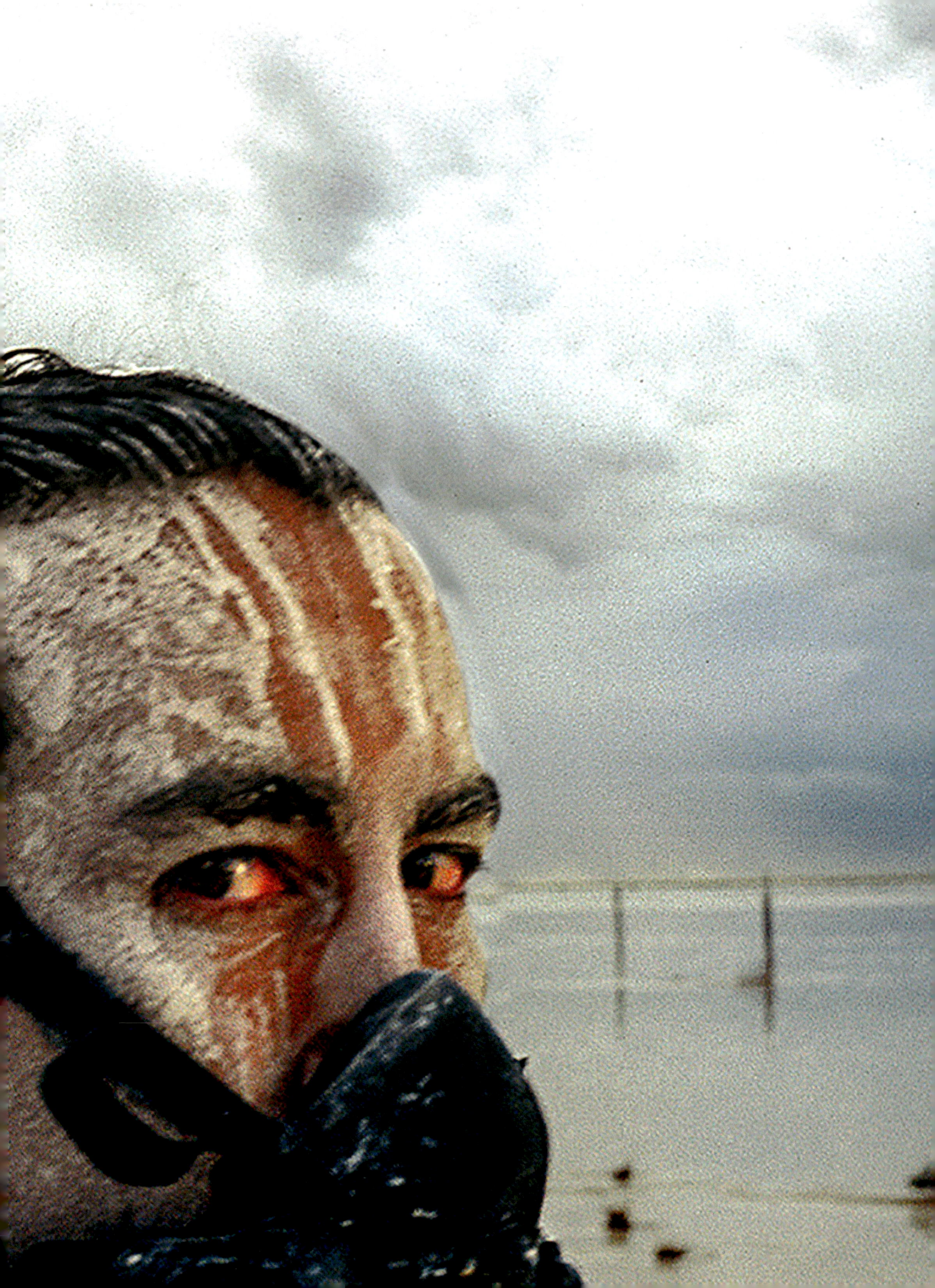

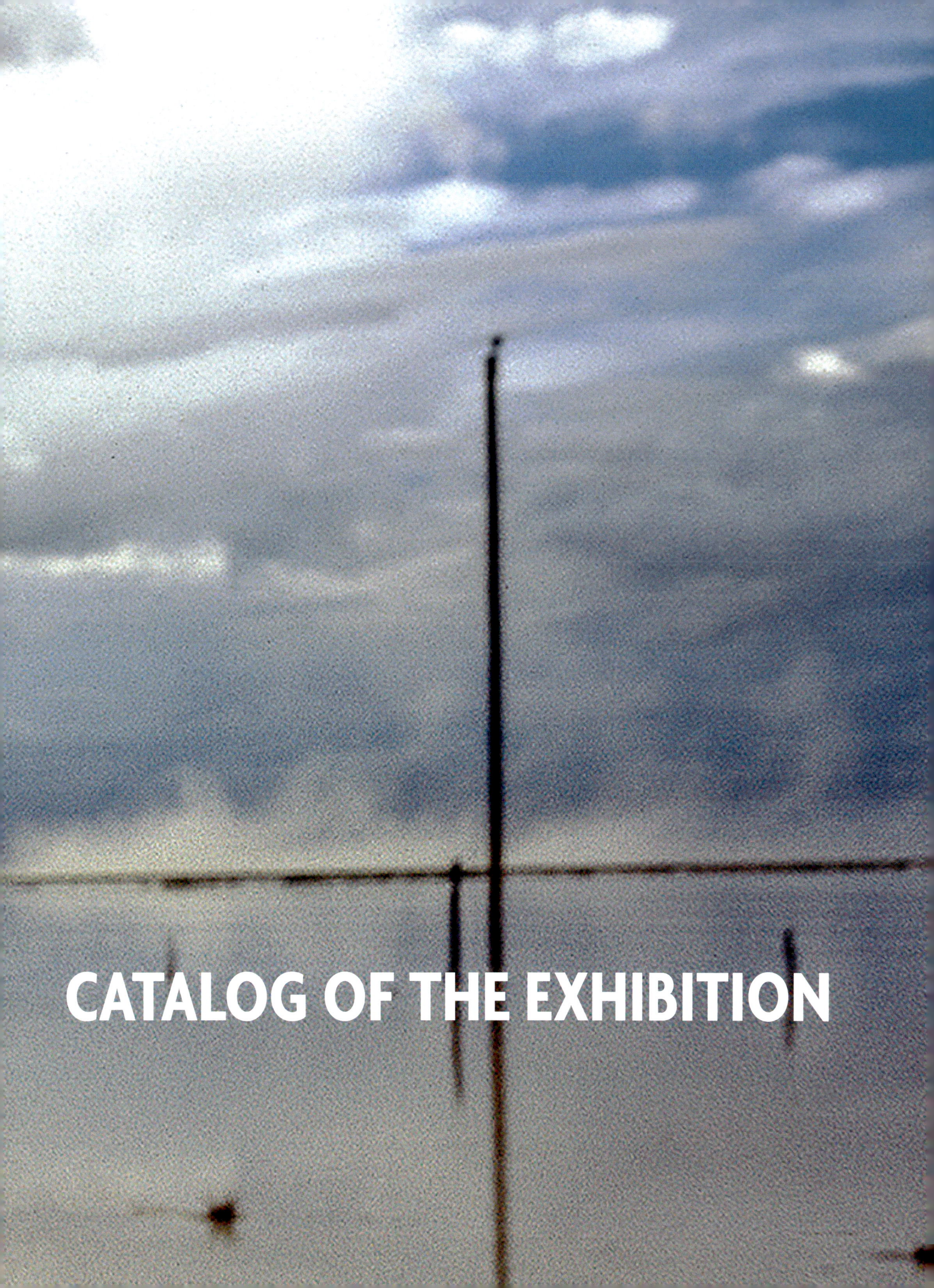

CATALOG OF THE EXHIBITION

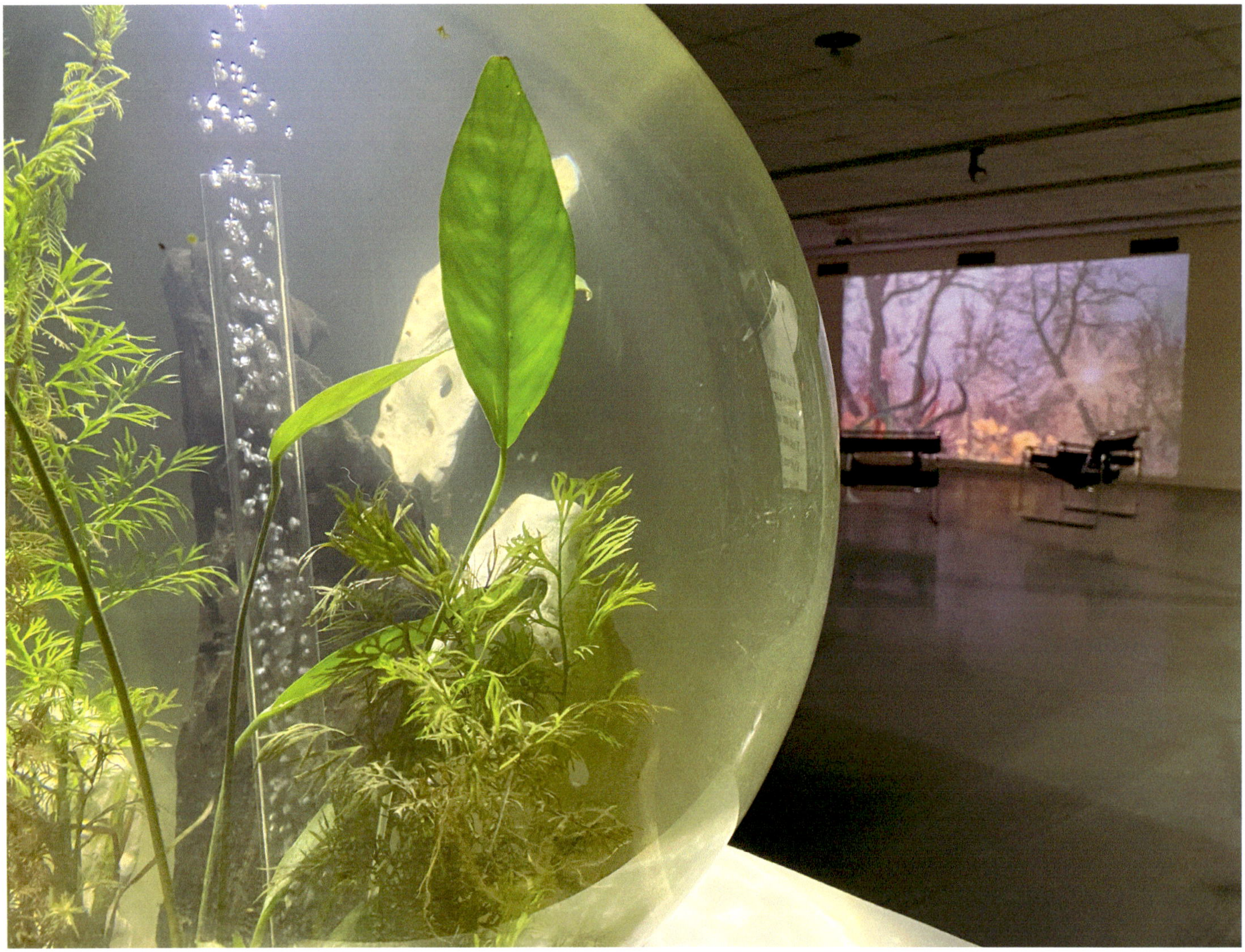

CORAL PROJECTS, THE FIRST FIVE YEARS: 2016 TO 2021, 2021
with the participation of John Torreano, Locus (Thale Fastvold and Tanja Thorjussen), Rev. Houston R. Cypress of Love the Everglades, Jeremy Olson, Gabriela Vainsencher, Donald Fortescue, Ebony G. Patterson, Waichung Wu, and Rachel Frank
Site-specific installation, dimensions variable. HD video with audio, inkjet prints, aquarium, ceramics, driftwood, aquatic plants, sketches and site-plans
Courtesy of Vanessa Albury/Coral Projects

VANESSA ALBURY

The installation *Coral Projects* is focused on a group's efforts to re-wild oyster and coral populations throughout the world, with proposed and actualized underwater exhibitions. Albury, with other artists, has collaborated on various iterations of *Coral Projects*. In this new *Coral Project*, she and artist Rachel Frank amplify each other's work. This Project also includes Frank's *Mini Mangrove Kernos Ceramic Vessels* placed in the Everglades by Albury and then photographed by Frank. In these sculptures, she creates evocative interpretations of still lifes, reminiscent of the Everglades' mud formations. Also, sketches by both Albury and Frank propose an "ocean friendly carbon-sink material" to rebuild coral and oyster life in waterways. Albury creates floating screens from the debris of plants in the Everglades, where the shadows and light we see in water dissolve. A film by Thale Fastvold and Tanja Thorjussen from Locus, an Oslo-based, artist collective, shows the performance of a "reverent honey ceremony." Other highlights are *Burned Out Film Aquarium*, 2019, an aquarium populated by freshwater plants, rocks, driftwood and ceramics that presents both the fragility of coral reefs and oysters in waterways and how artists address this crisis with work that impacts our desire to foster nature's regrowth and regeneration.

VANESSA ALBURY

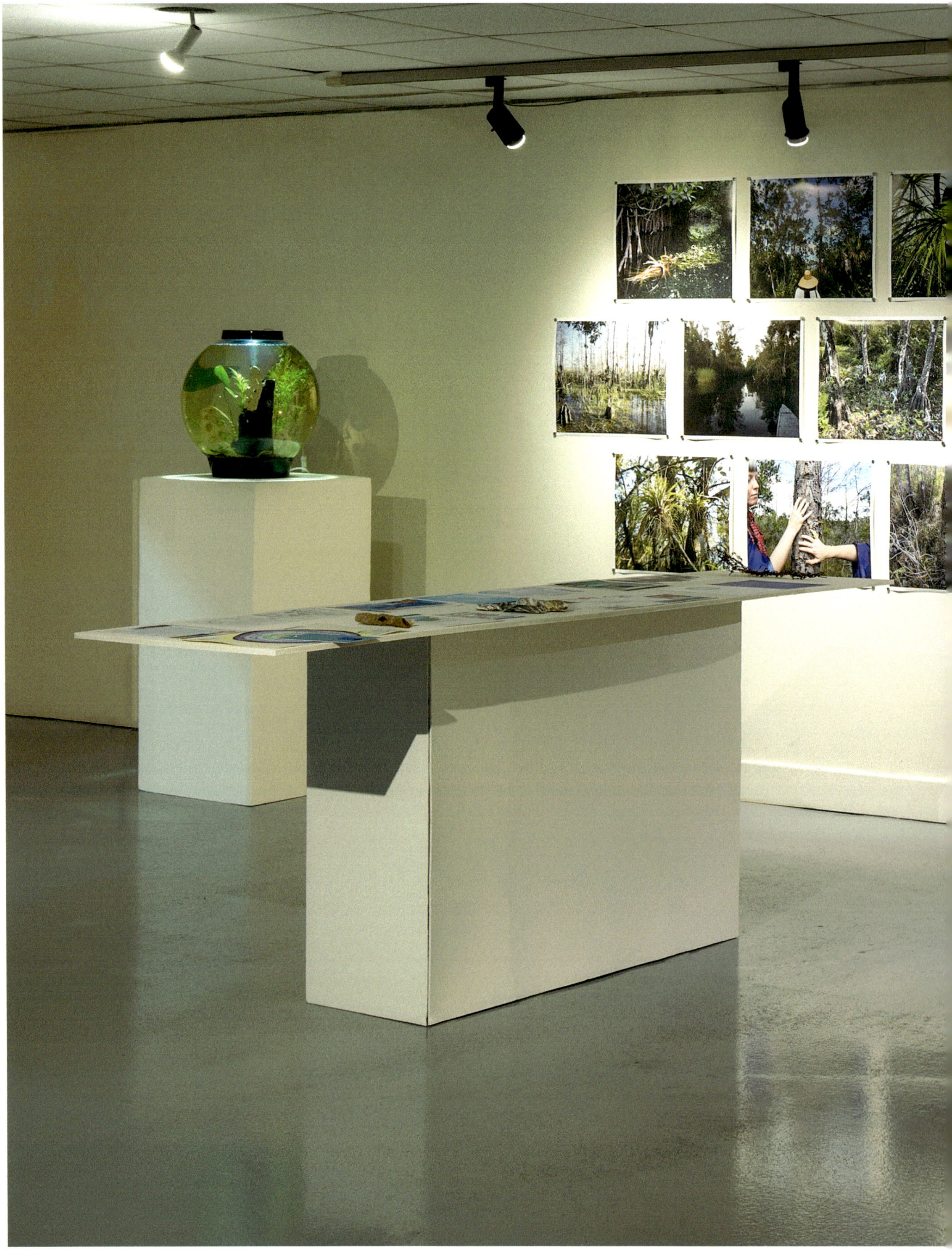

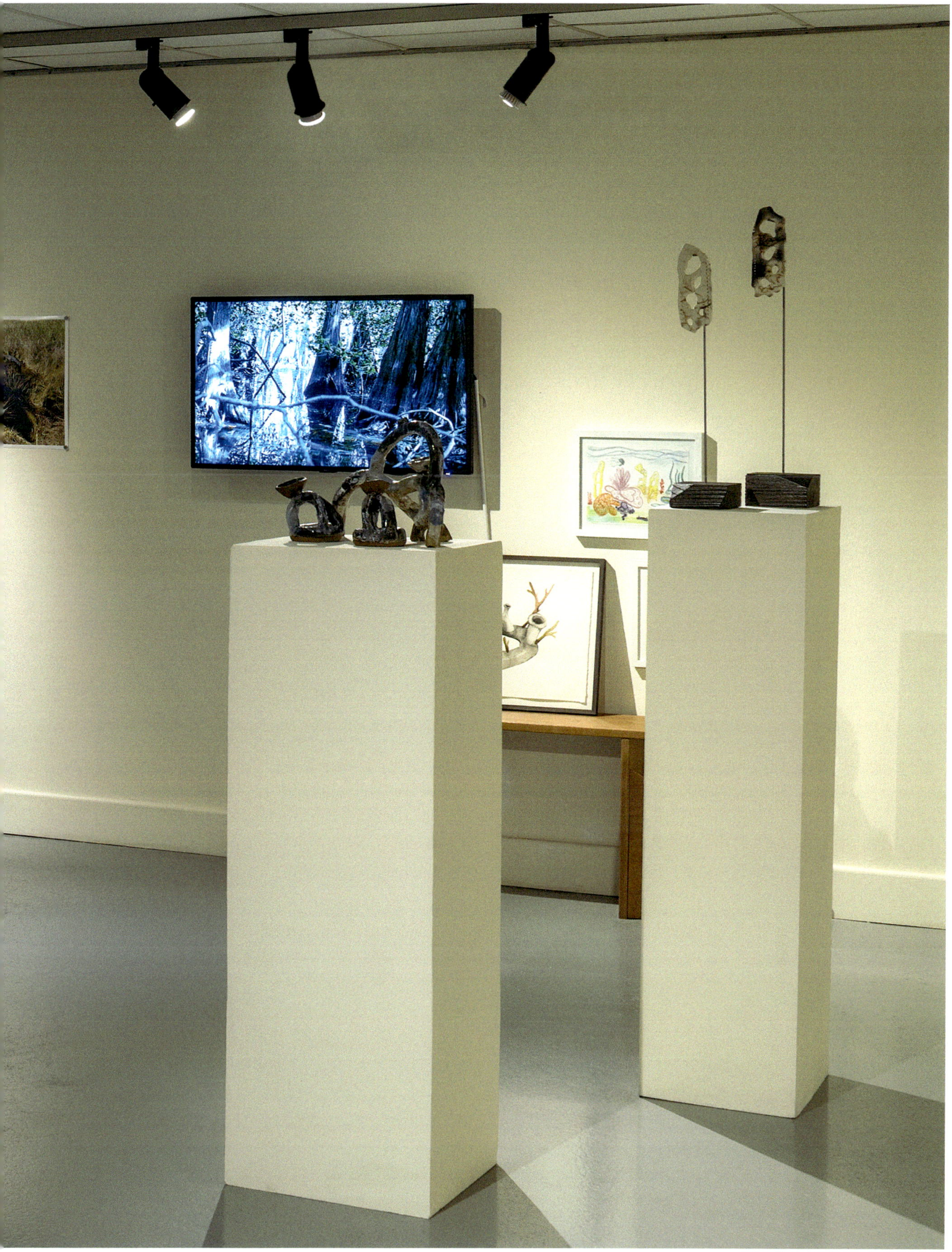

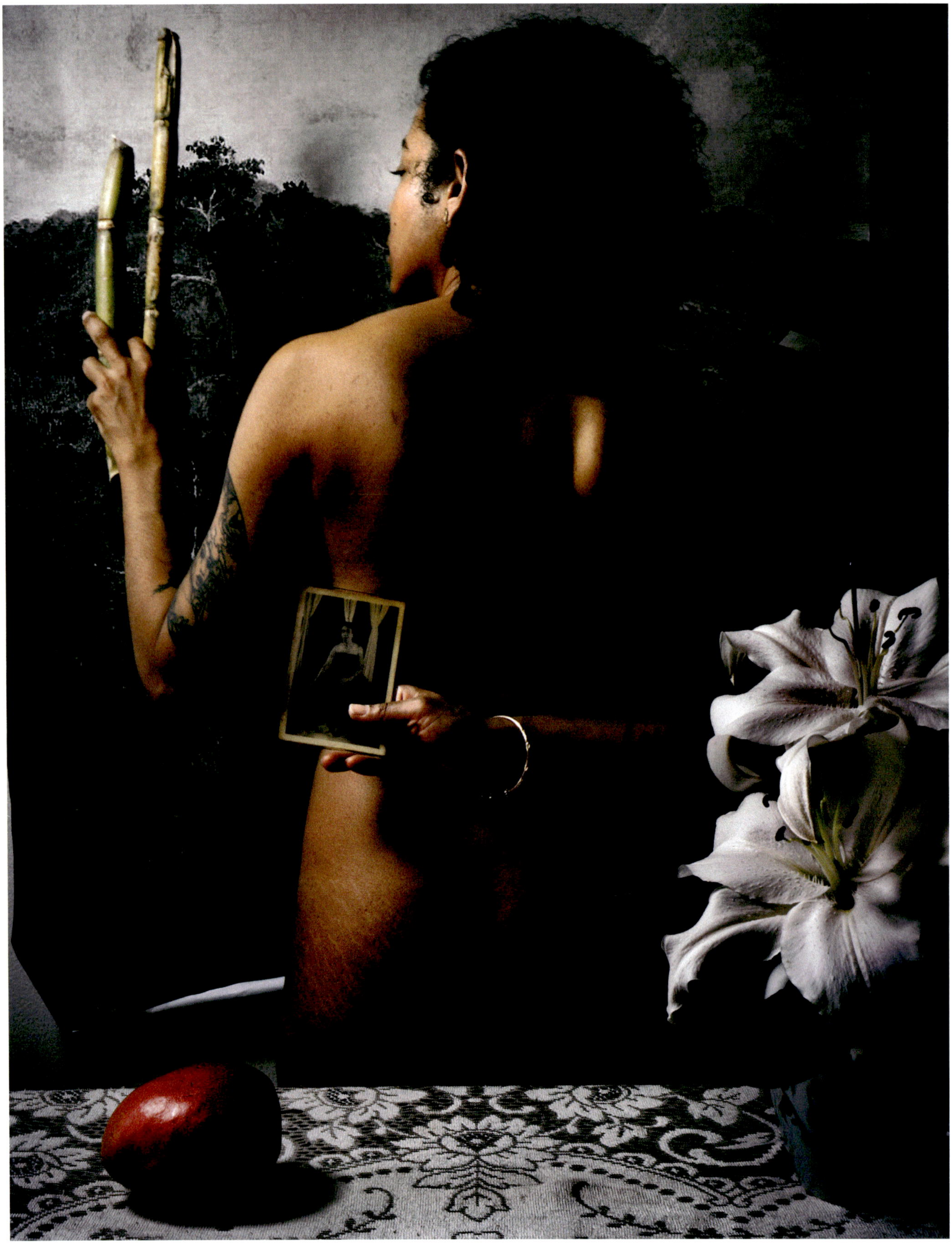

WITH JEWELRY, SUGARCANE AND GRANDMA'S PHOTOGRAPH, 2019
Photograph, 11 x 14 inches
Courtesy of the artist

SAMANTHA BOX

In *Portable Homeland #6*, Box addresses themes of accessibility and justice in a Diaspora Herbarium, an "herbarium of resistance." Her photographs and the herbarium pose the question: Can the Caribbean diaspora communities in the Bronx harness the healing quality and nurture of "ethno medicine" — a knowledge of plants that heal, which is held by enslaved and indigenous peoples. Her work delves into questions of sustainability, pondering both access to these plants and the ensurance of the plants' survival — an act of resistance to colonial domination. The *Portable Homeland* installation is a physical manifestation of ancestral food and its use in the diaspora. Lemongrass, ginger, papaya, and turmeric are embedded within the soil, while Box fuses gem-like colored photographs to the structure to create both translucency and luminosity that wobble under the glow of a grow light, its color spectrum encouraging a plant to grow. Within this composition, the artist encloses a portrait of her youthful self, which the viewer sees in abstracted flowing shapes. *Portable Homeland* becomes, then, a poignant homage to Box's Caribbean heritage, to the complexity of colonial influence within diaspora communities, and to ongoing efforts to sustain the traditions of healing foods.

SAMANTHA BOX

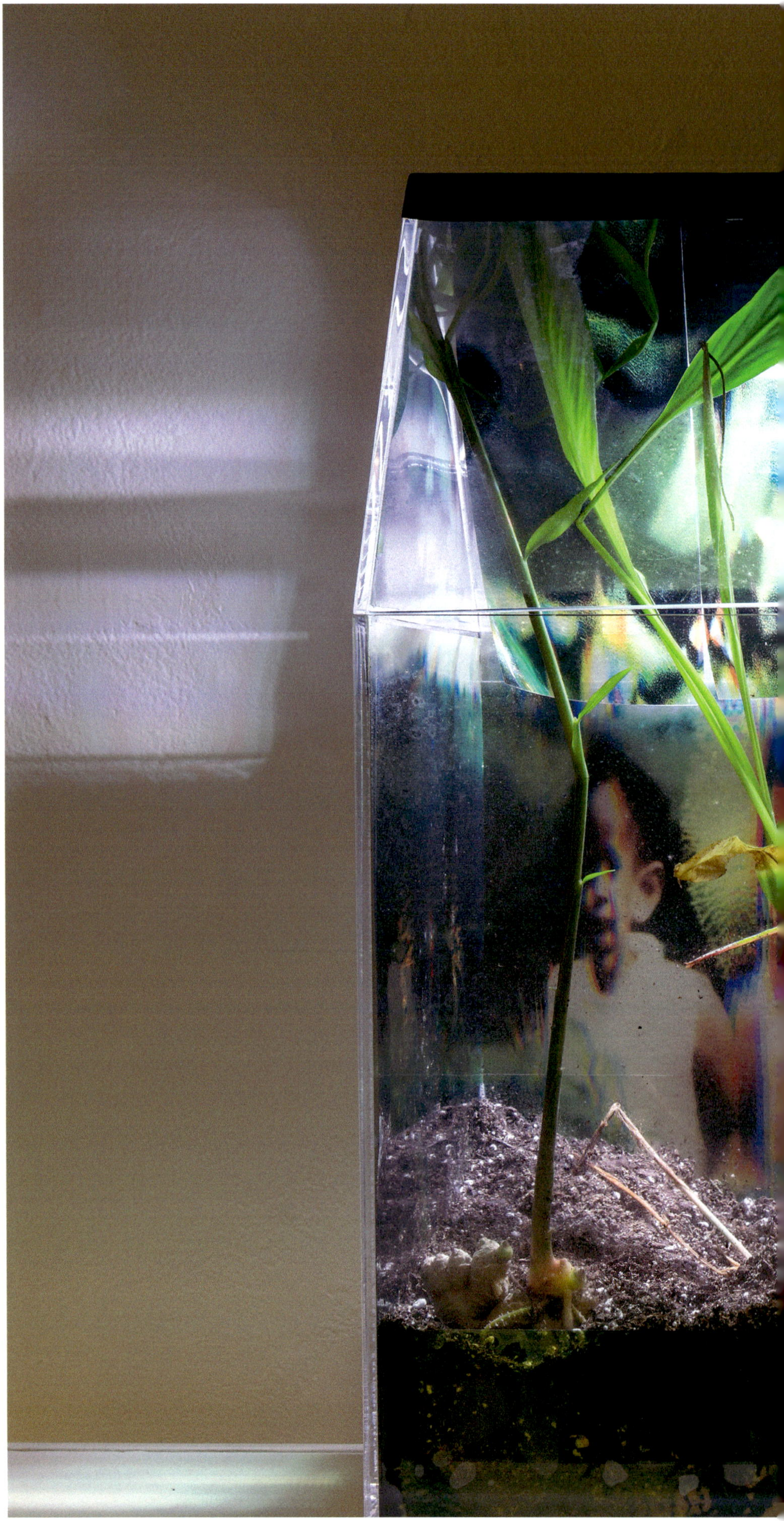

PORTABLE HOMELAND #6,
2021
Mixed media herbarium installation,
Dimensions variable
Courtesy of the artist

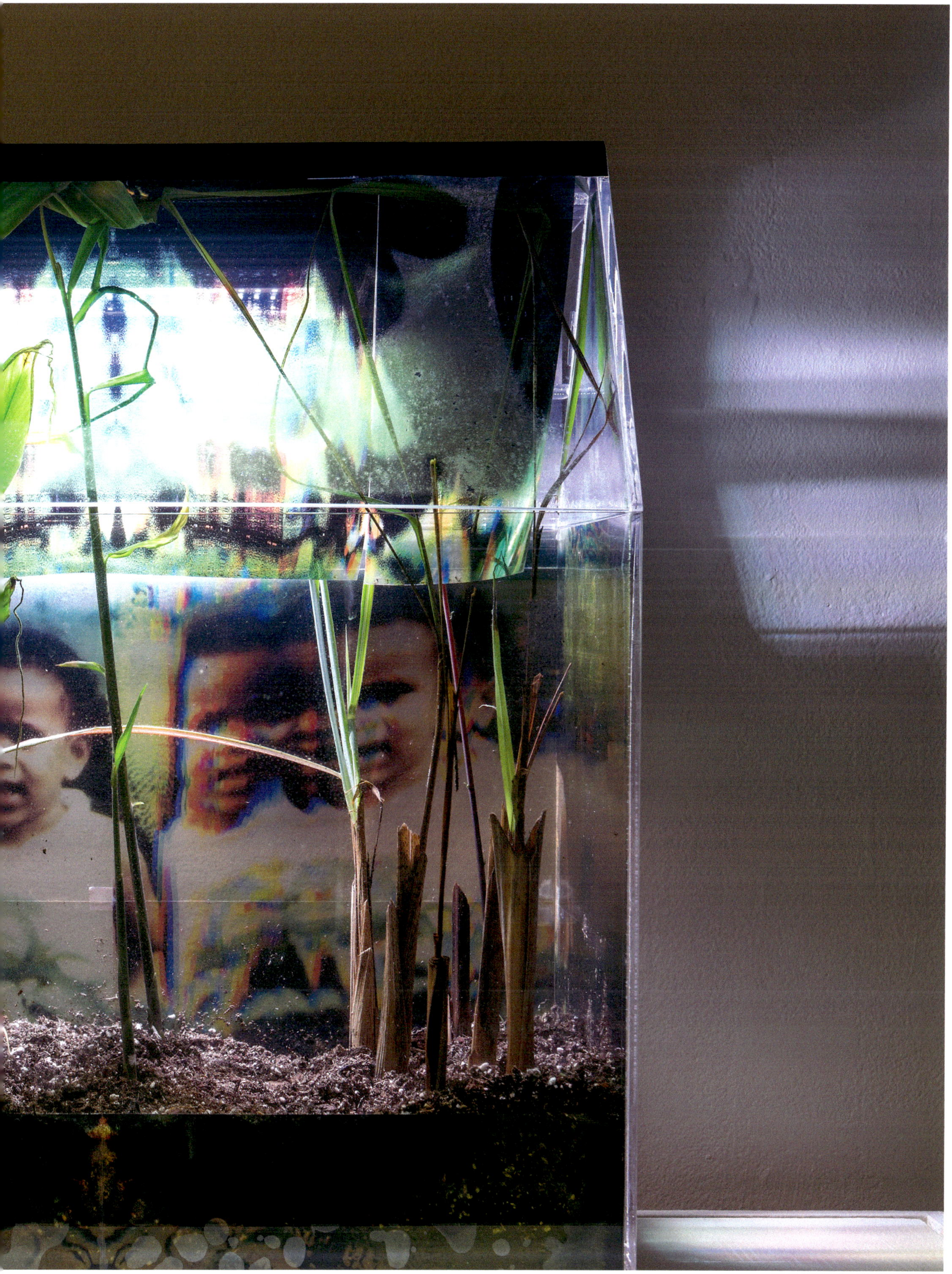

SAMANTHA BOX

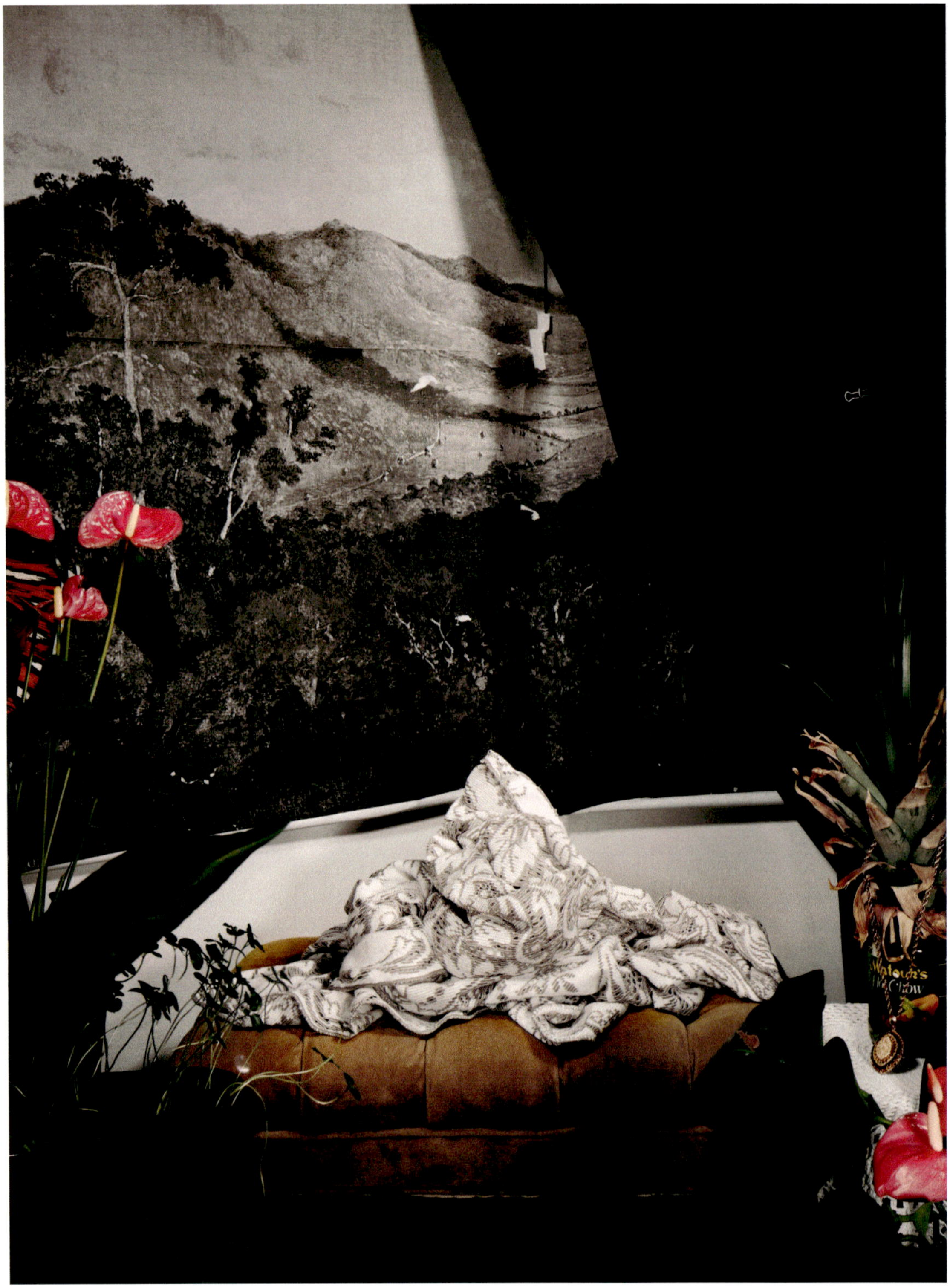

COLLAPSE, 2020
Photograph, 20 x 24 inches
Courtesy of the artist

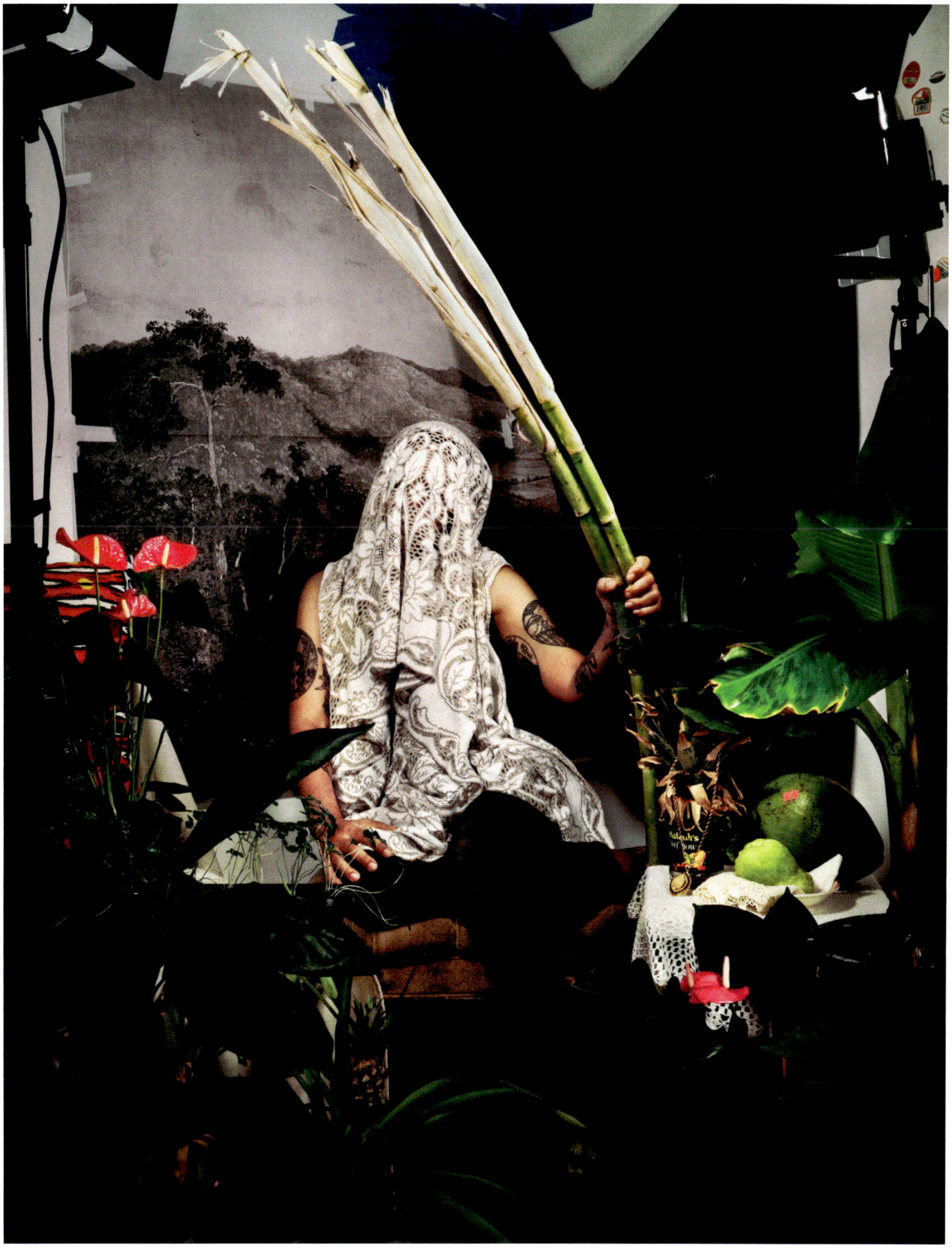

EDGES, 2020
Photograph, 20 x 24 inches
Courtesy of the artist

SAMANTHA BOX

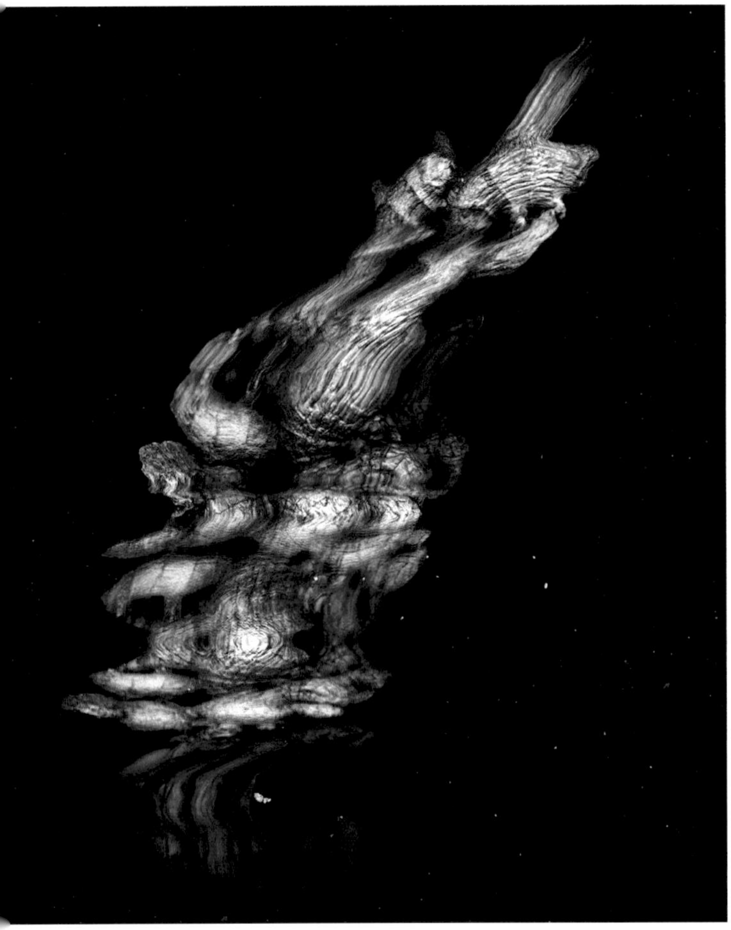

UNTITLED 1, 2021
Archival inkjet print
20 x 24 inches
Courtesy of the artist

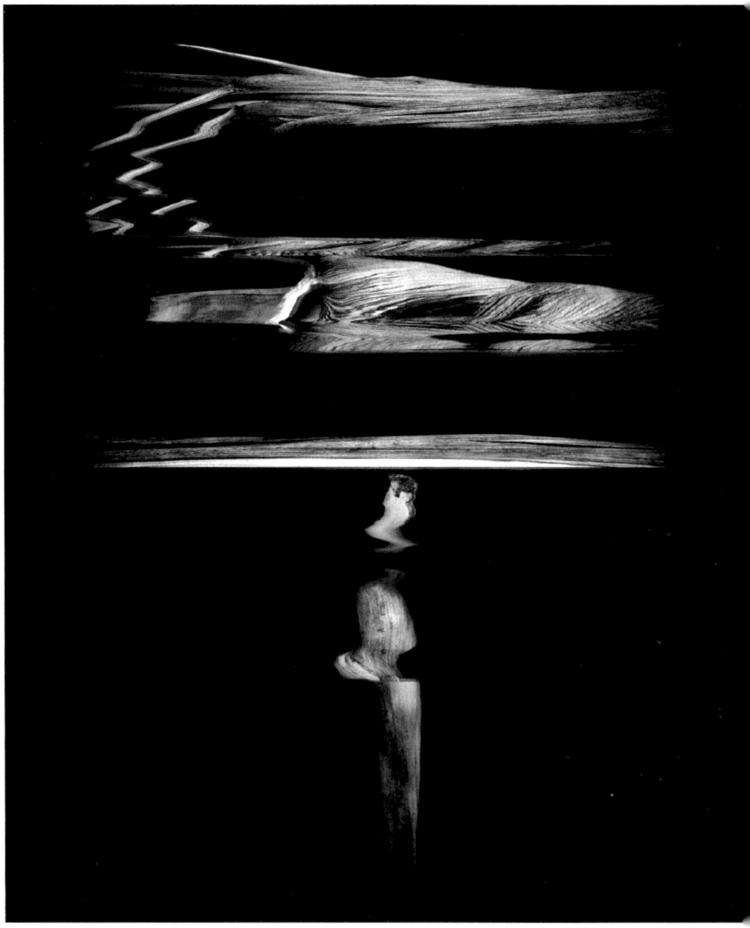

UNTITLED 2A, 2021
Archival inkjet print
20x24 inches
Courtesy of the artist

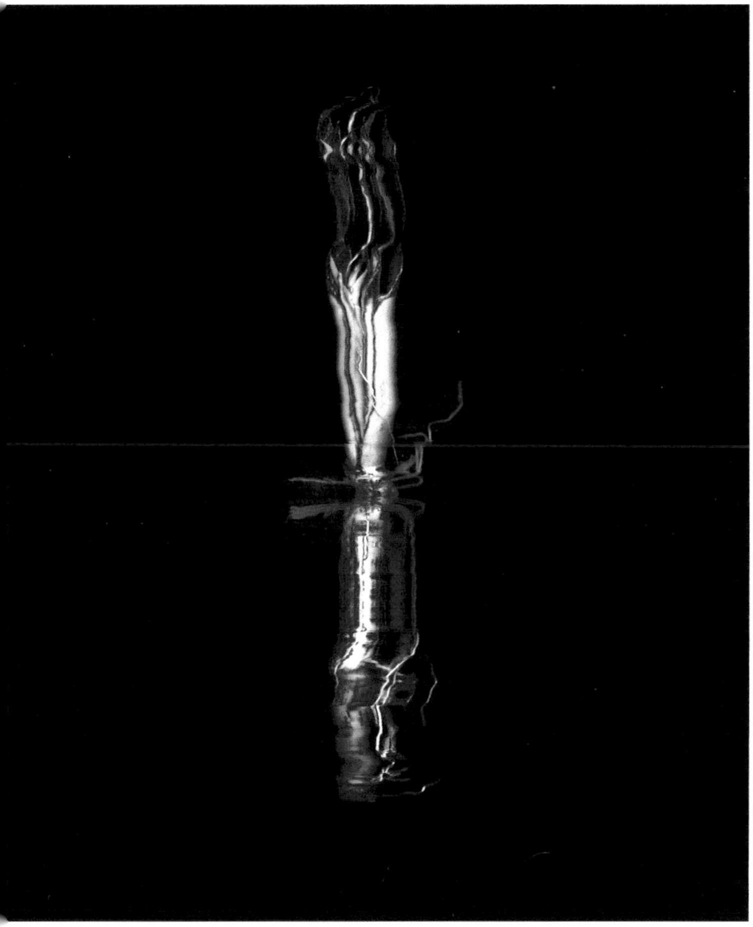

UNTITLED 2B, 2021
Archival inkjet print
20 x 24 inches
Courtesy of the artist

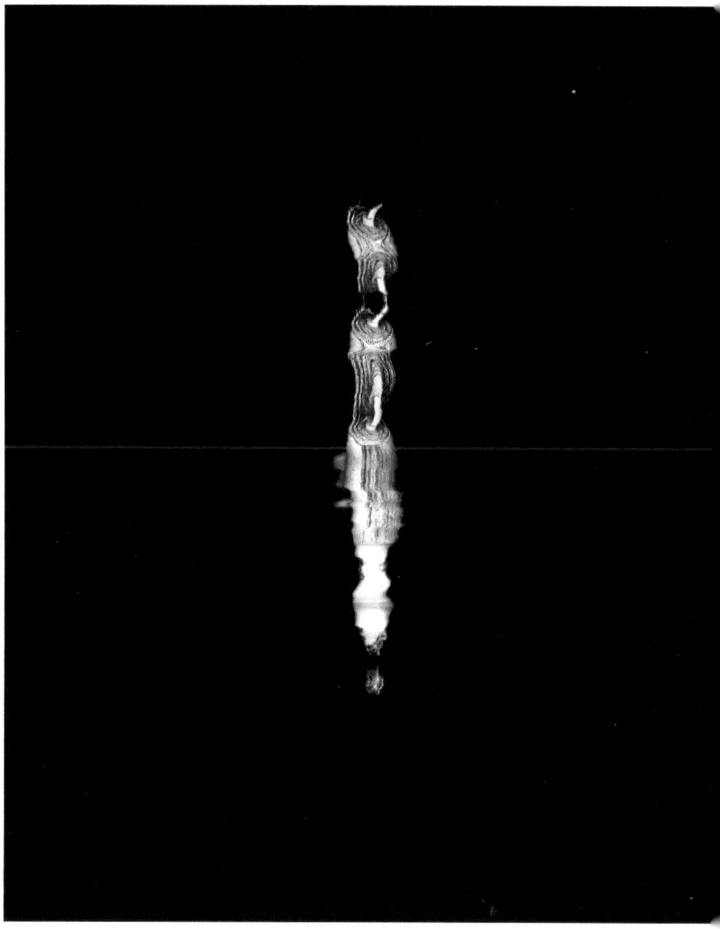

UNTITLED 2C, 2021
Archival inkjet print
20 x 24 inches
Courtesy of the artist

SAMANTHA BOX

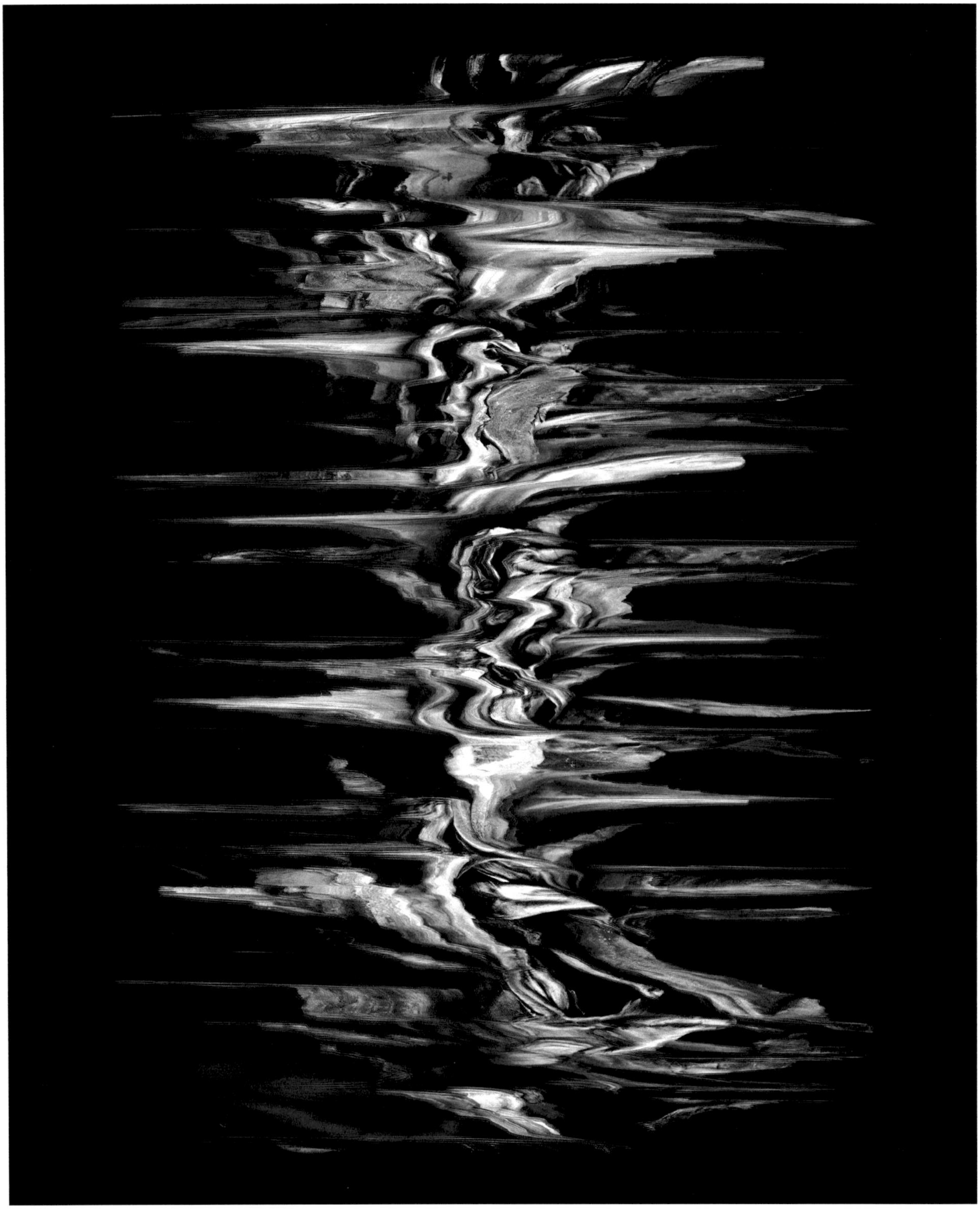

UNTITLED 3, 2021
Archival inkjet print
20 x 24 inches
Courtesy of the artist

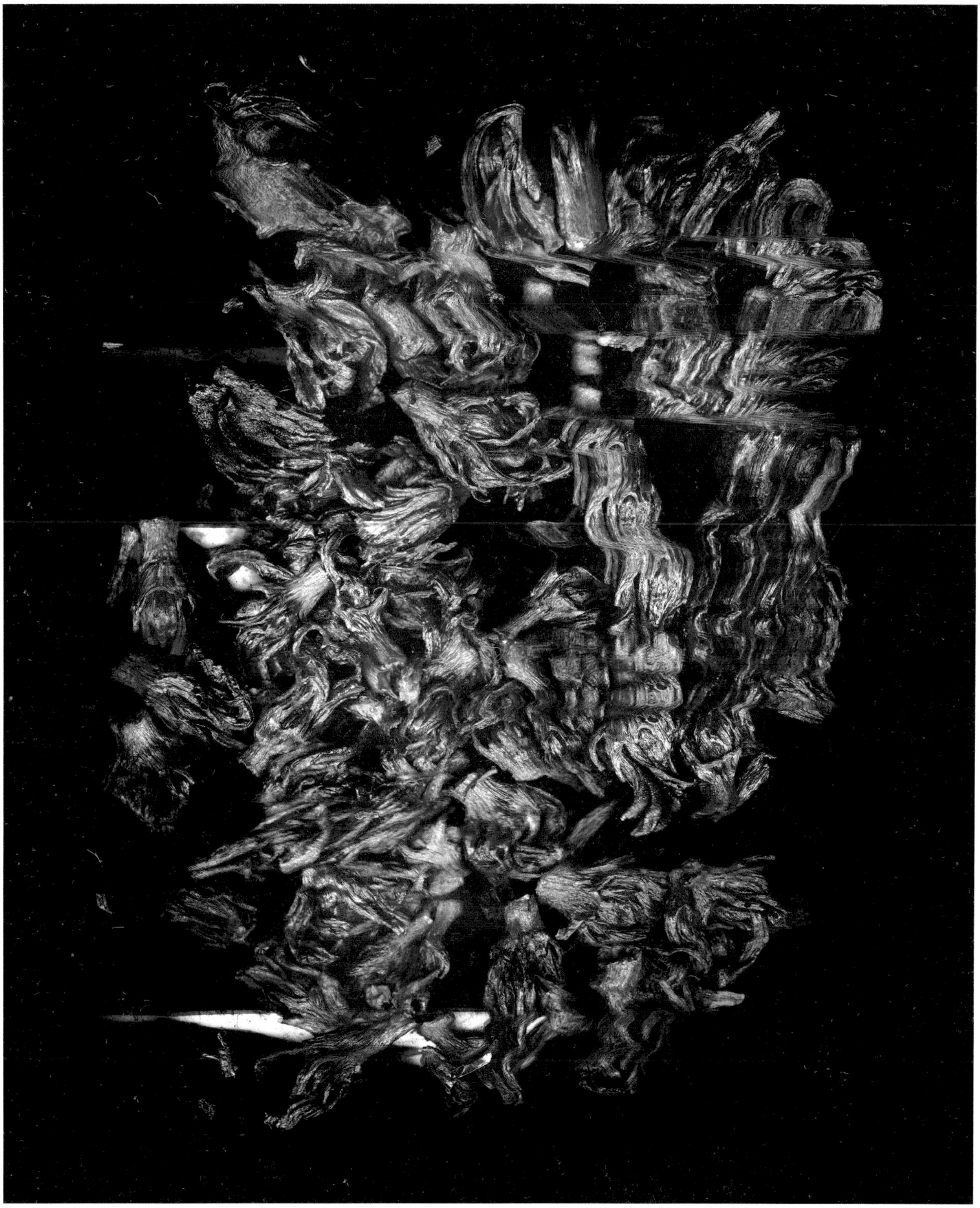

UNTITLED 4, 2021
Archival inkjet print
20 x 24 inches
Courtesy of the artist

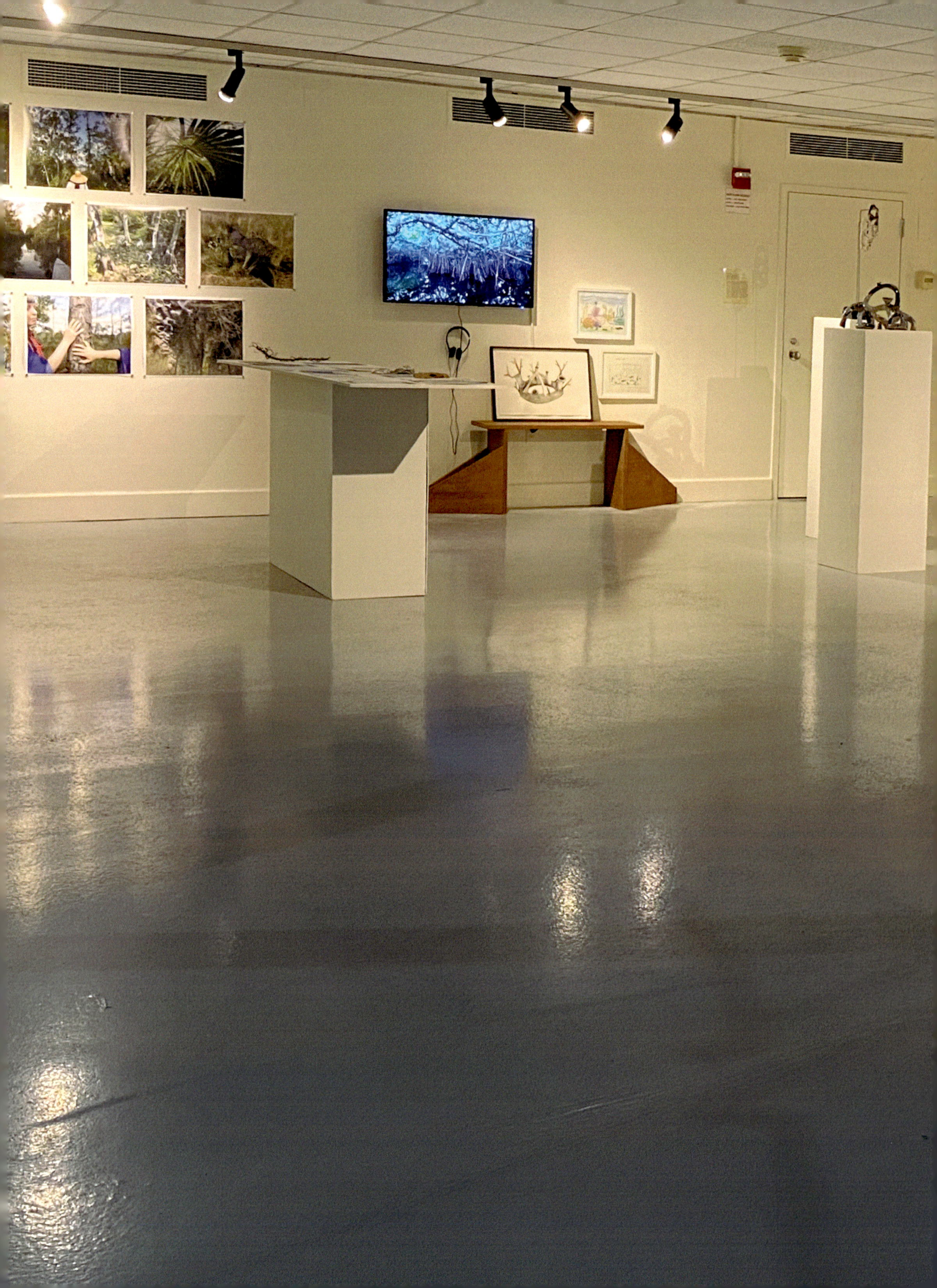

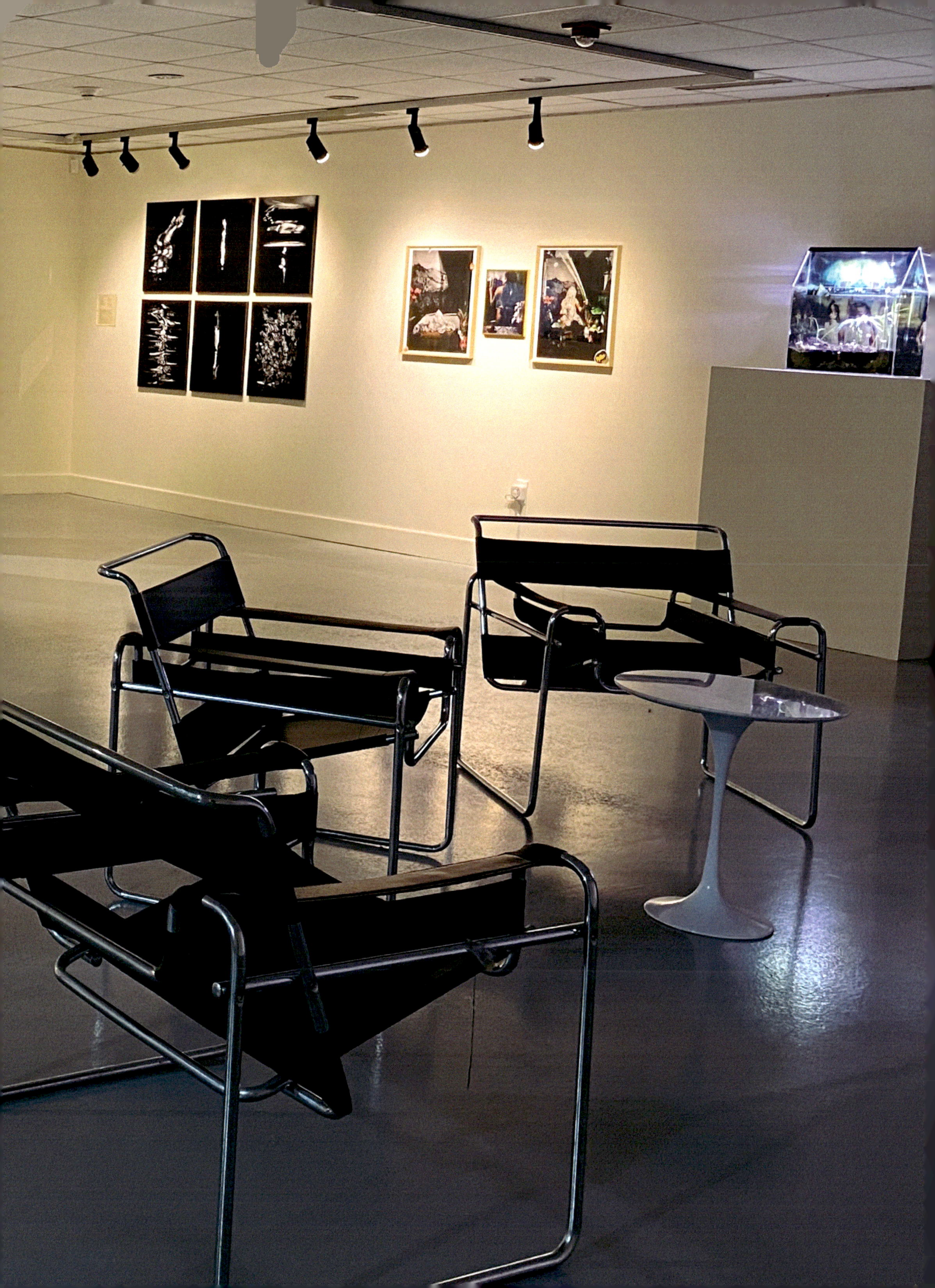

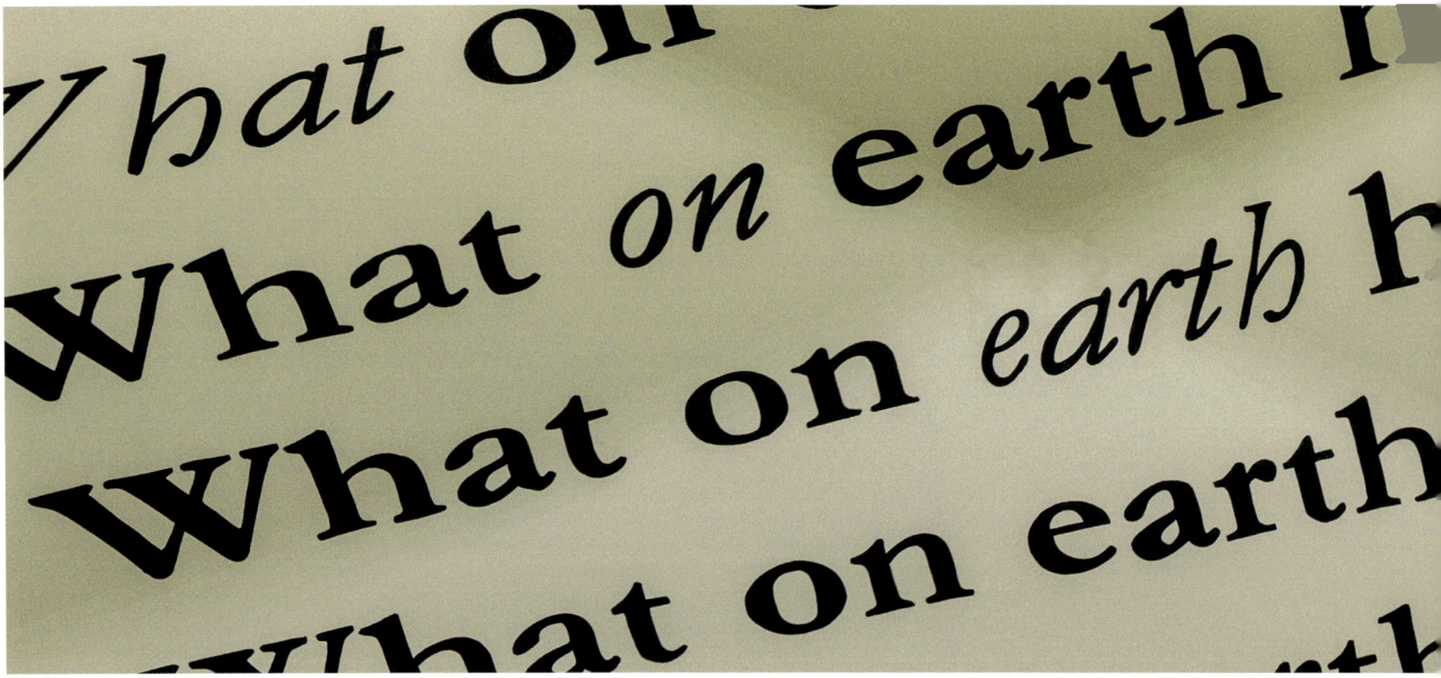

WHAT ON EARTH? (HAVE YOU DONE), 2012/2021
Detail. Mixed media, site-specific installation
Dimensions variable
Courtesy of the artist

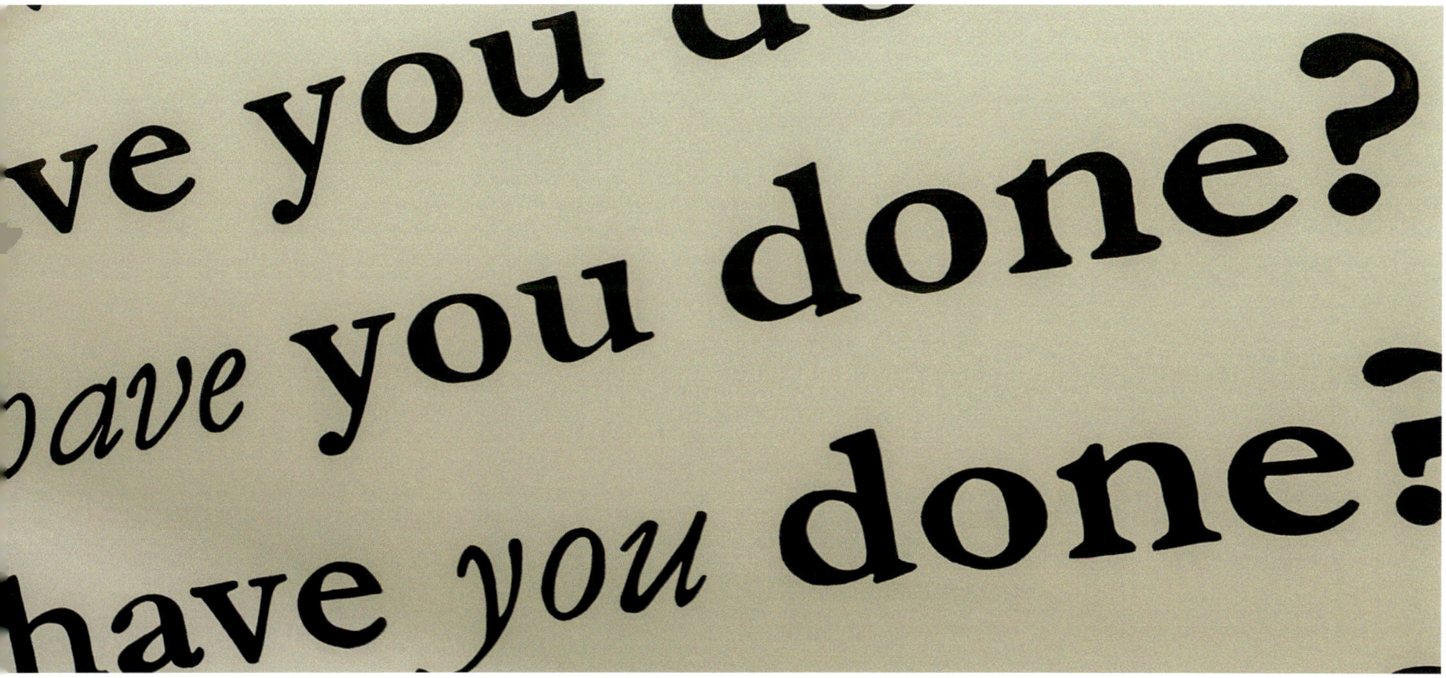

NICKY ENRIGHT

Enright's striking black-and-white word installation stops viewers cold. The question "What on earth have you done?" is repeated six times on two facing walls. Each iteration of the question, with its changing emphasis, seems to implicate viewers as to their role on the planet as well as to the meaning they give to their lives. Here Enright installed the text dramatically, cropped and at a diagonal, so it maximizes the curve of Lehman Gallery's distinctive Lobby Rotunda, while simultaneously appearing to sink into the ground and rise off the wall. A level, smaller version of this text-based installation appeared at the entrance to the Wave Hill iteration of *Eco-Urgency: Now or Never*, expanding on a work from 2012 that originally appeared in the book *25 Great Sentences and How They Got That Way* by Geraldine Woods. Enright said, "I modify the emphasis of each individual word to interrogate the relationships among visual text, aural sound, and literal meaning. Although the phrase remains the same, it is tweaked by italics to subtly change the meaning of the question/accusation. The viewer is encouraged to read and 'hear' the words differently in each line, thereby changing the significance of the expression and ultimately questioning individual accountability for the climate crisis."

NICKY ENRIGHT

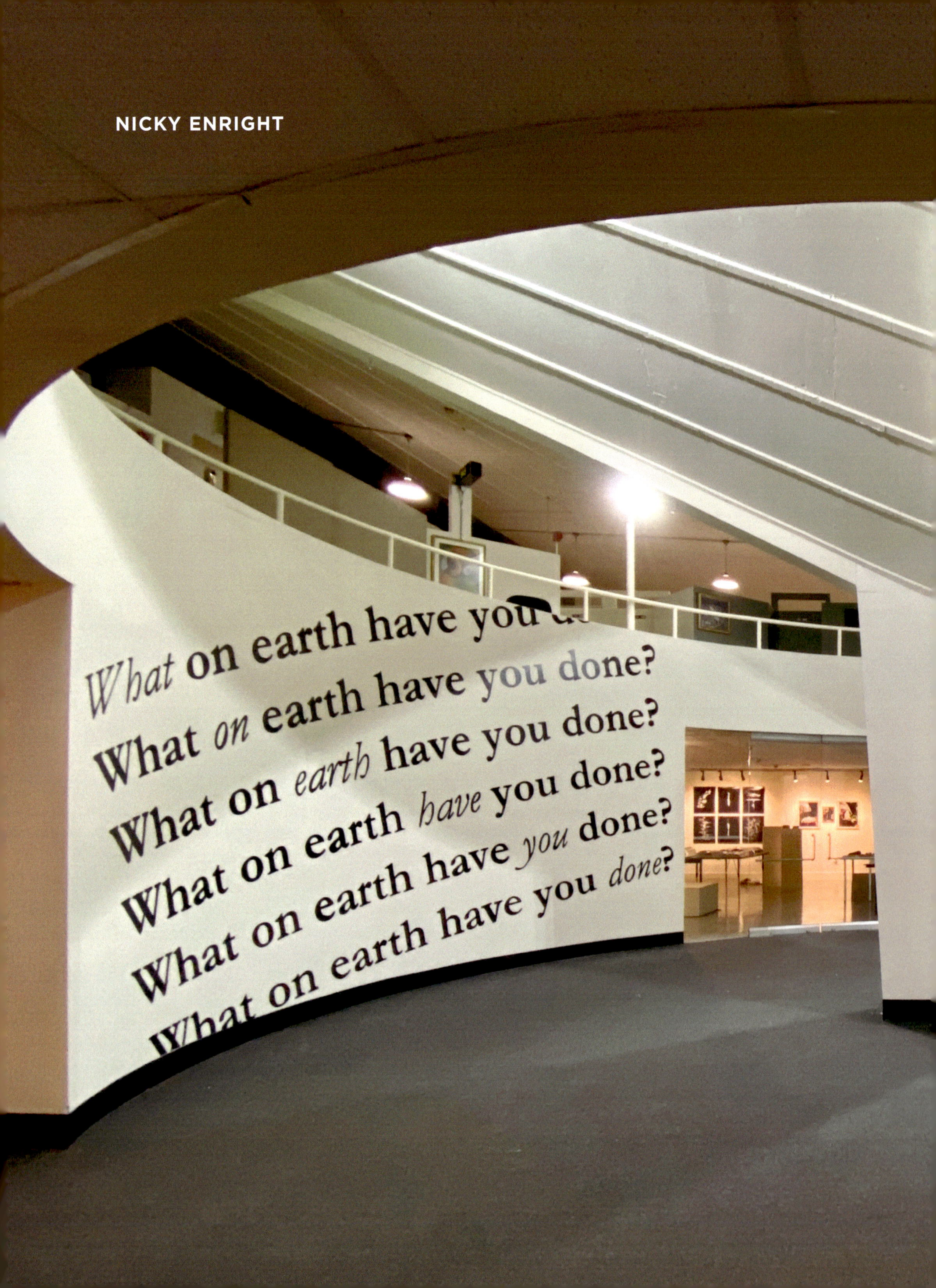

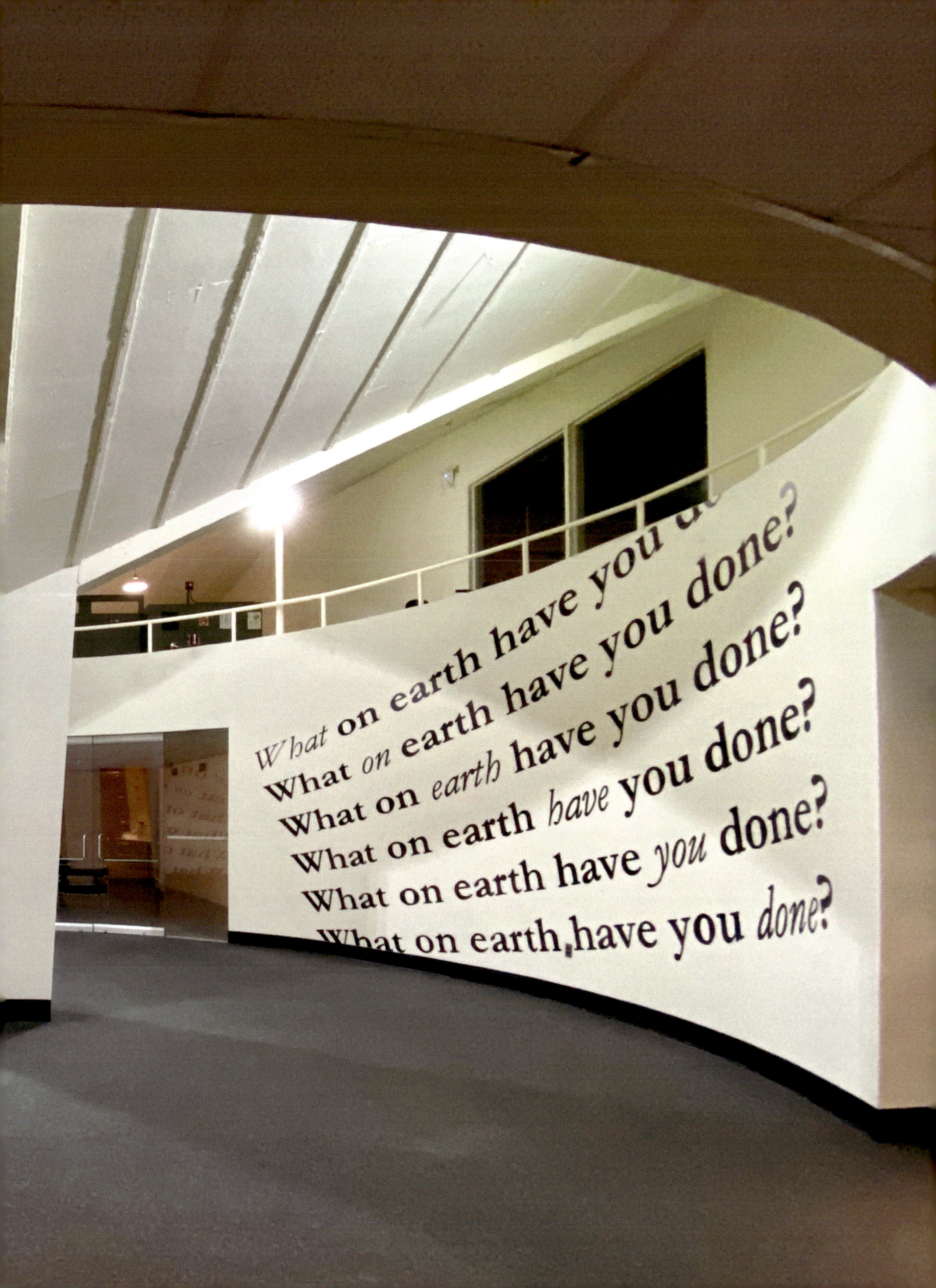
What on earth have you
What on earth have you done?
What on earth have you done?
What on earth have you done?
What on earth have you done?
What on earth have you done?

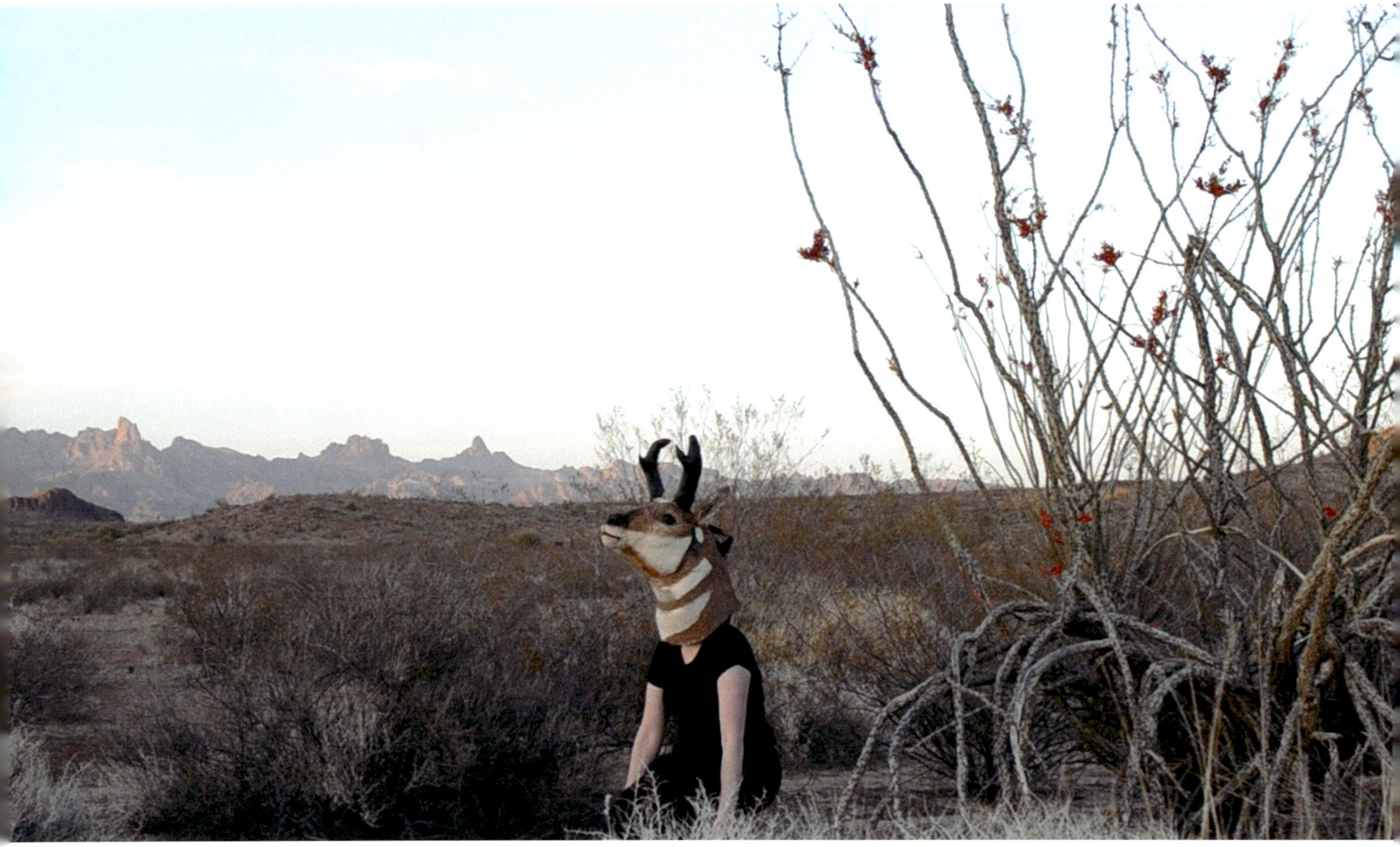

THRESHOLDS, 2019
Single-channel HD video with sculptural objects
4:19 minutes
Courtesy of Rachel Frank Studio

RACHEL FRANK

Frank's video *Thresholds* reflects on "wildlife corridors," the pathways for wildlife-linking habitats that are disrupted by man-made growth. Through her explorations and contemplations of the borderlands of Arizona's southern Sonoran Desert, Frank addresses wildlife fragmentation; migration; borders; climate change; droughts; and the changing uses of the desert. She dons her large-scale sculptures of animal heads in performances, using her body to interact with the landscape and to draw our attention to the practice of "re-wilding" — reintroducing a species into its natural environment in an effort to restore wild eco-systems. She depicts this process, sometimes joyful, sometimes traumatic from the animal's perspective. The mournful gaze of her animals draw in the viewer as the animals appear to plea for help to right the climate crisis.

Sentinel Offering Lekythos: Clapper Rail and Mangrove, 2021, Frank's newest sculpture, depicts the Clapper Rail, a bird that is an "indicator species," and one of the first to show the effects of climate change. Frank cared for Clapper Rail birds and has released them into the environment in her wildlife rehabilitation work. The form of this sculpture is based on an ancient offering vessel associated with loss or funerary rites. The vessel that once held liquids like olive oil or wine is now engulfed by Frank's runny, sinuous shapes that echo the bird's habitat. The bird itself, perched atop the composition, appears in danger of melting. The artist completes the colors of this wetland habitat with gold leaf.

RACHEL FRANK

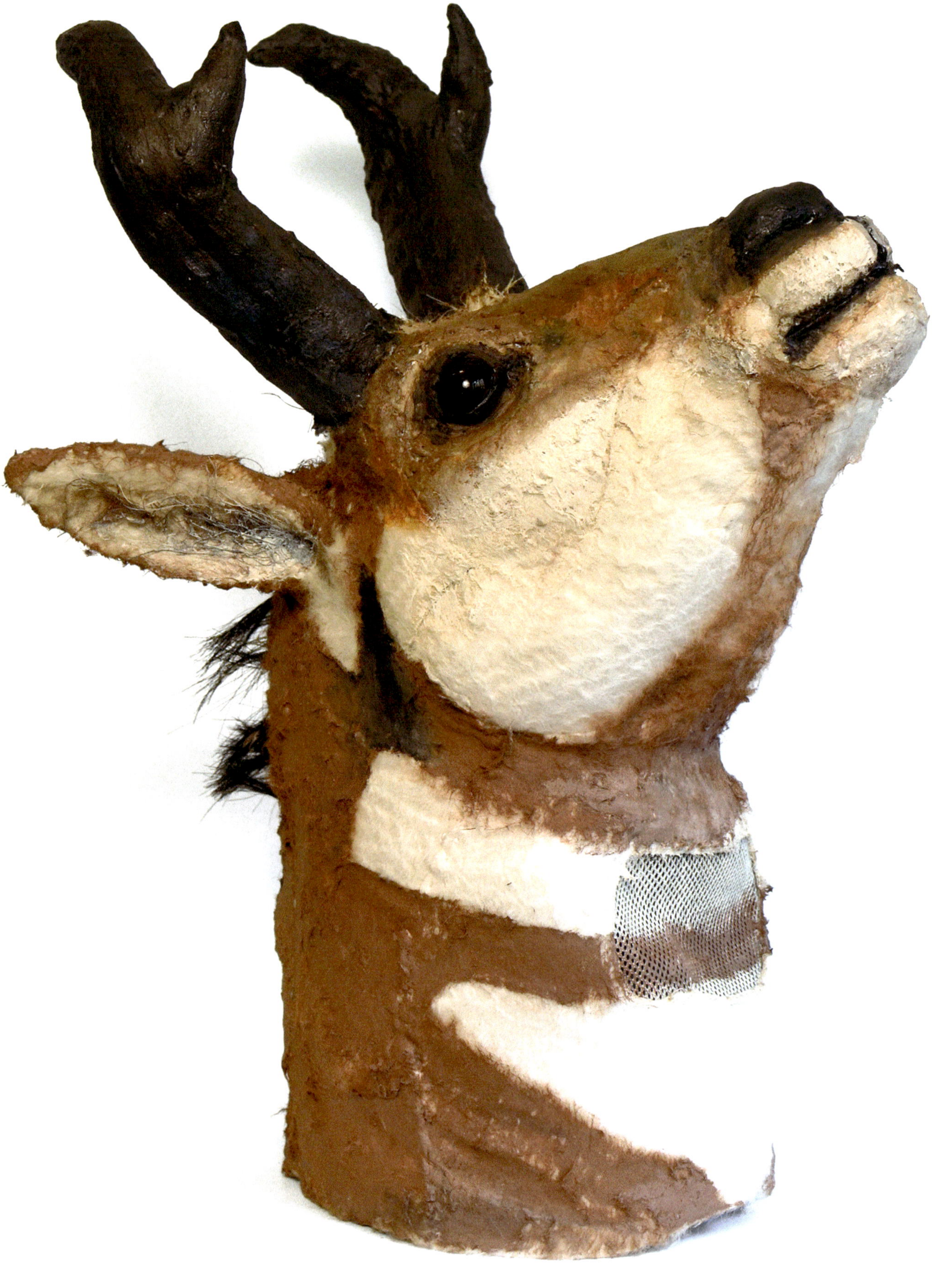

NORTH AMERICAN SONORAN PRONGHORN, 2017
Fabric, thread, acrylic,
and other materials
24 x 17½ x 13 inches
Courtesy of Rachel Frank Studio

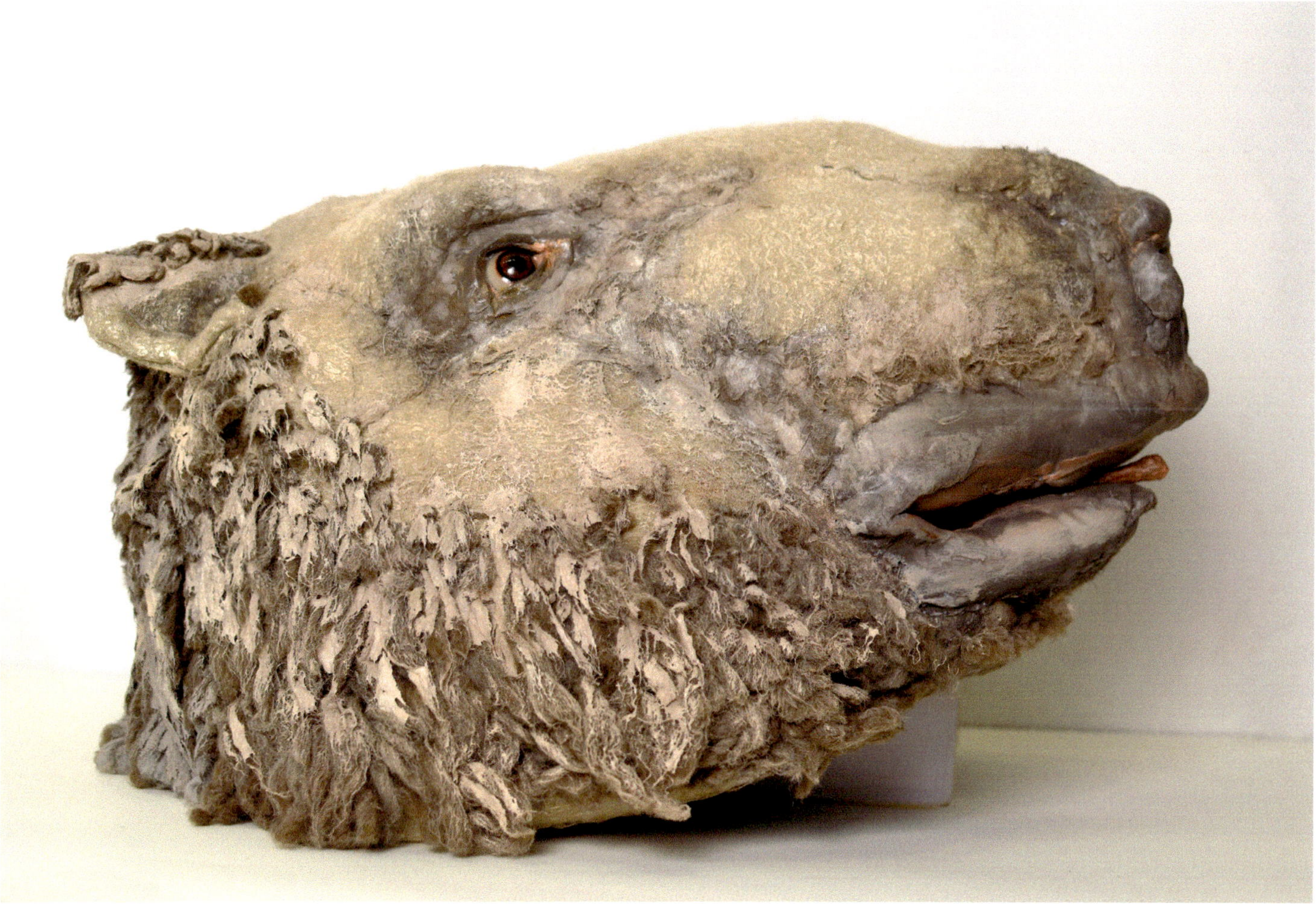

MYLODON, GROUND SLOTH, 2013
Acrylic paint, yarn, and other materials
22 x 36 x 9 inches
Courtesy of Rachel Frank Studio

RACHEL FRANK

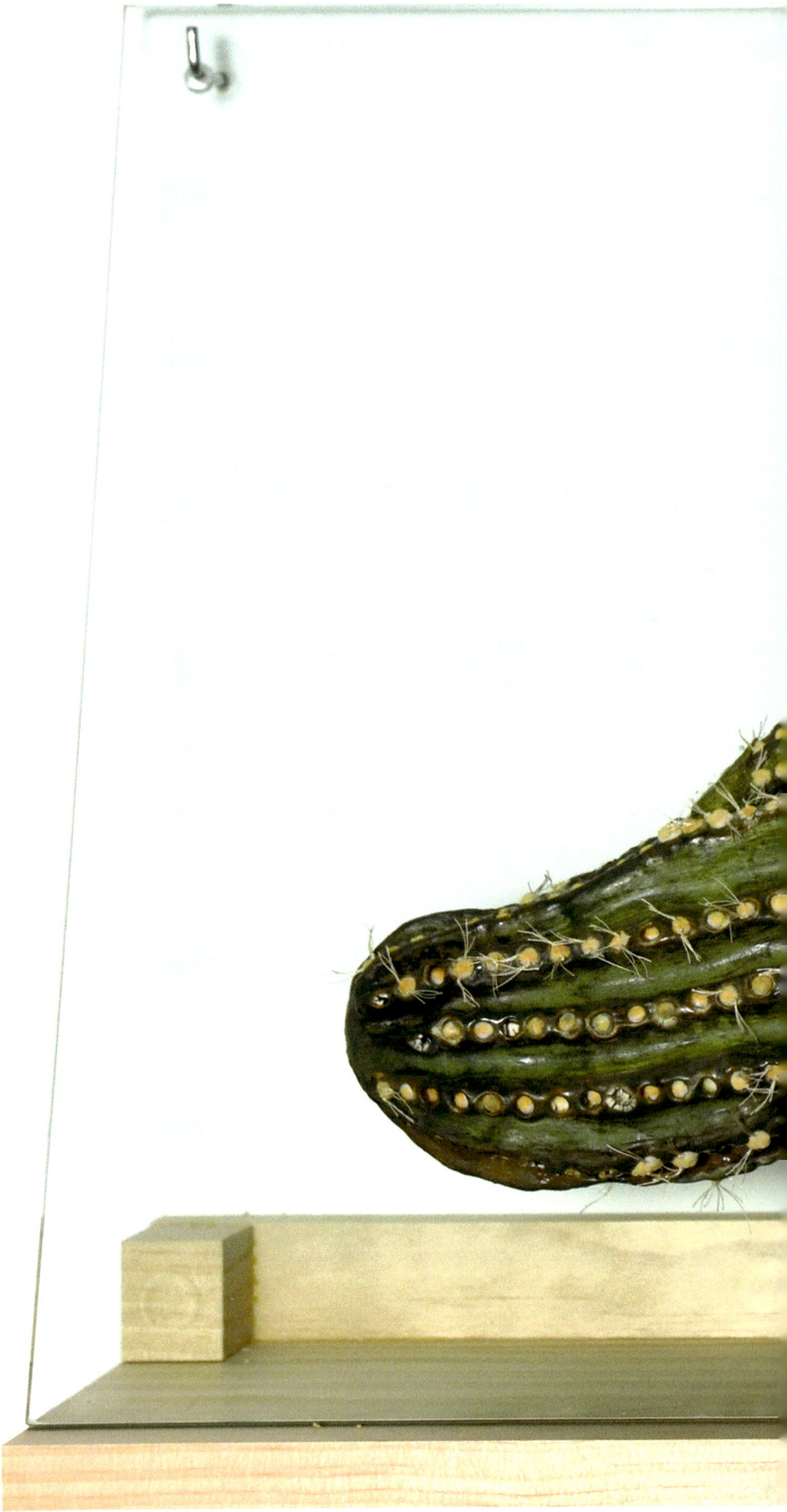

THRESHOLDS, SAGUARO ARM, 2018
Stoneware ceramic cactus with glazes, colored pencil, Plexiglas, glue, fabric, thread, wood, and hardware
18 x 24 x 7 inches
Courtesy of Rachel Frank Studio

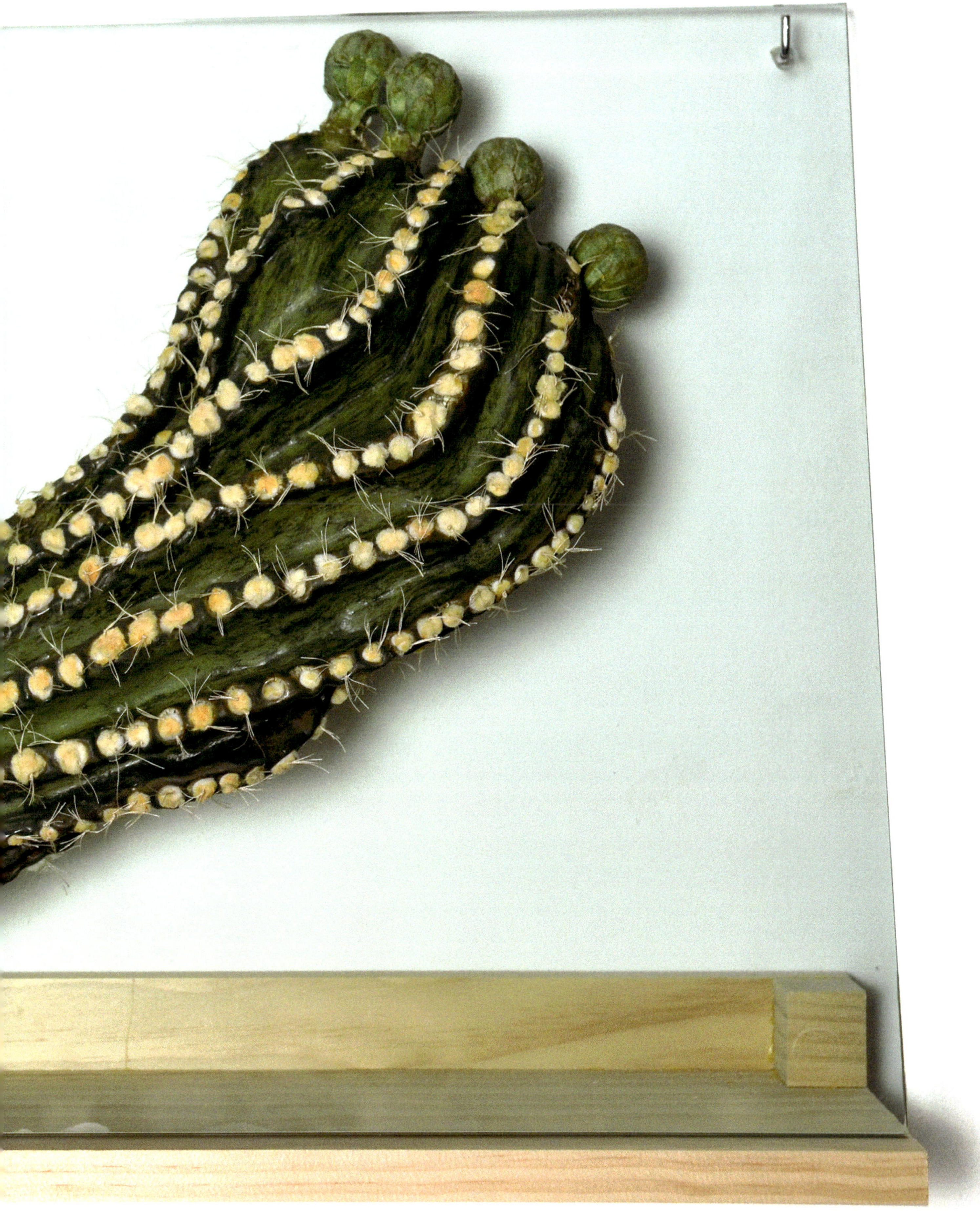

RACHEL FRANK

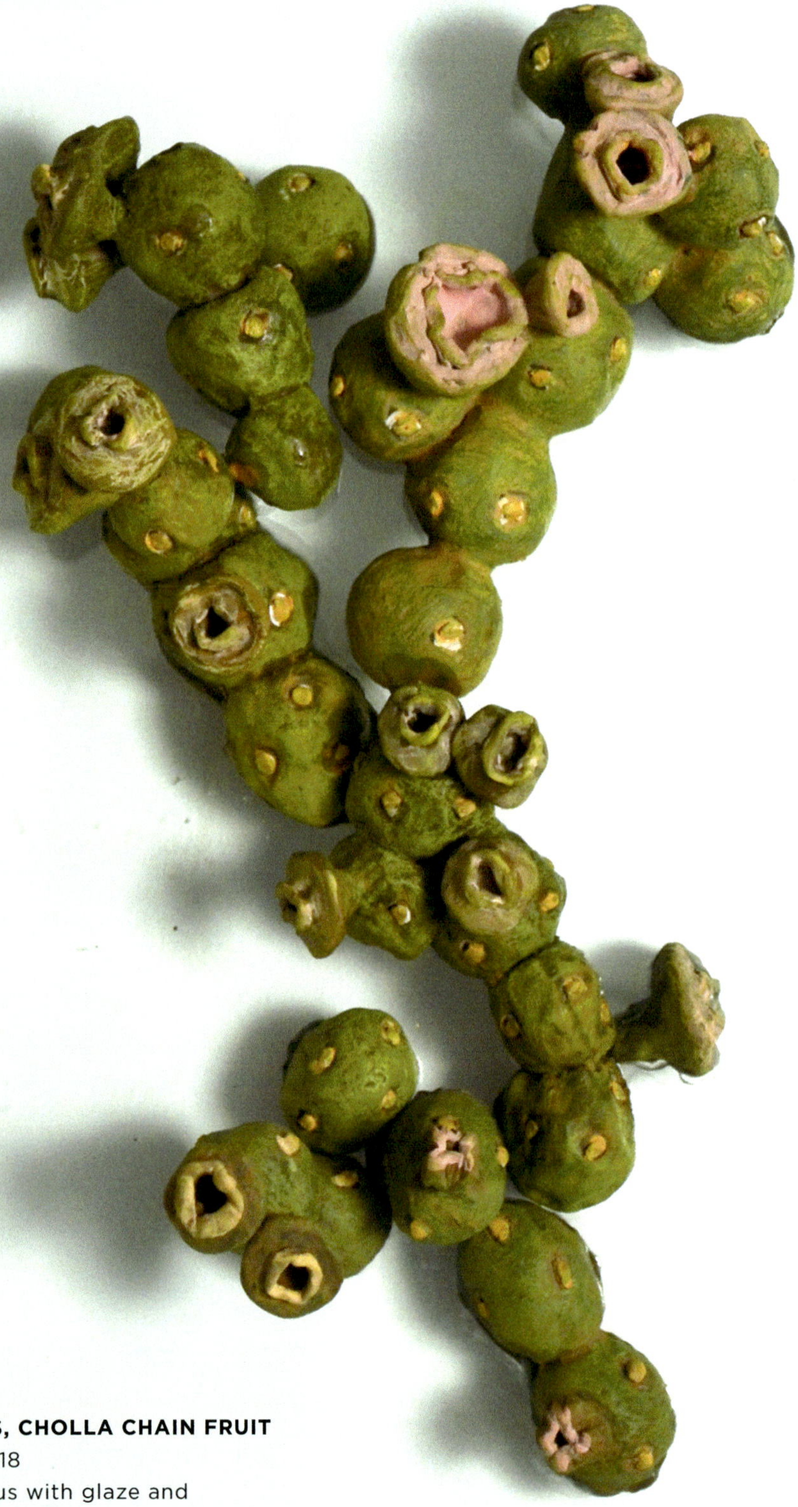

THRESHOLDS, CHOLLA CHAIN FRUIT CACTUS I, 2018
Ceramic cactus with glaze and colored pencil,
Plexiglas, glue, hardware
24 x 18 x 4 inches
Courtesy of Rachel Frank Studio

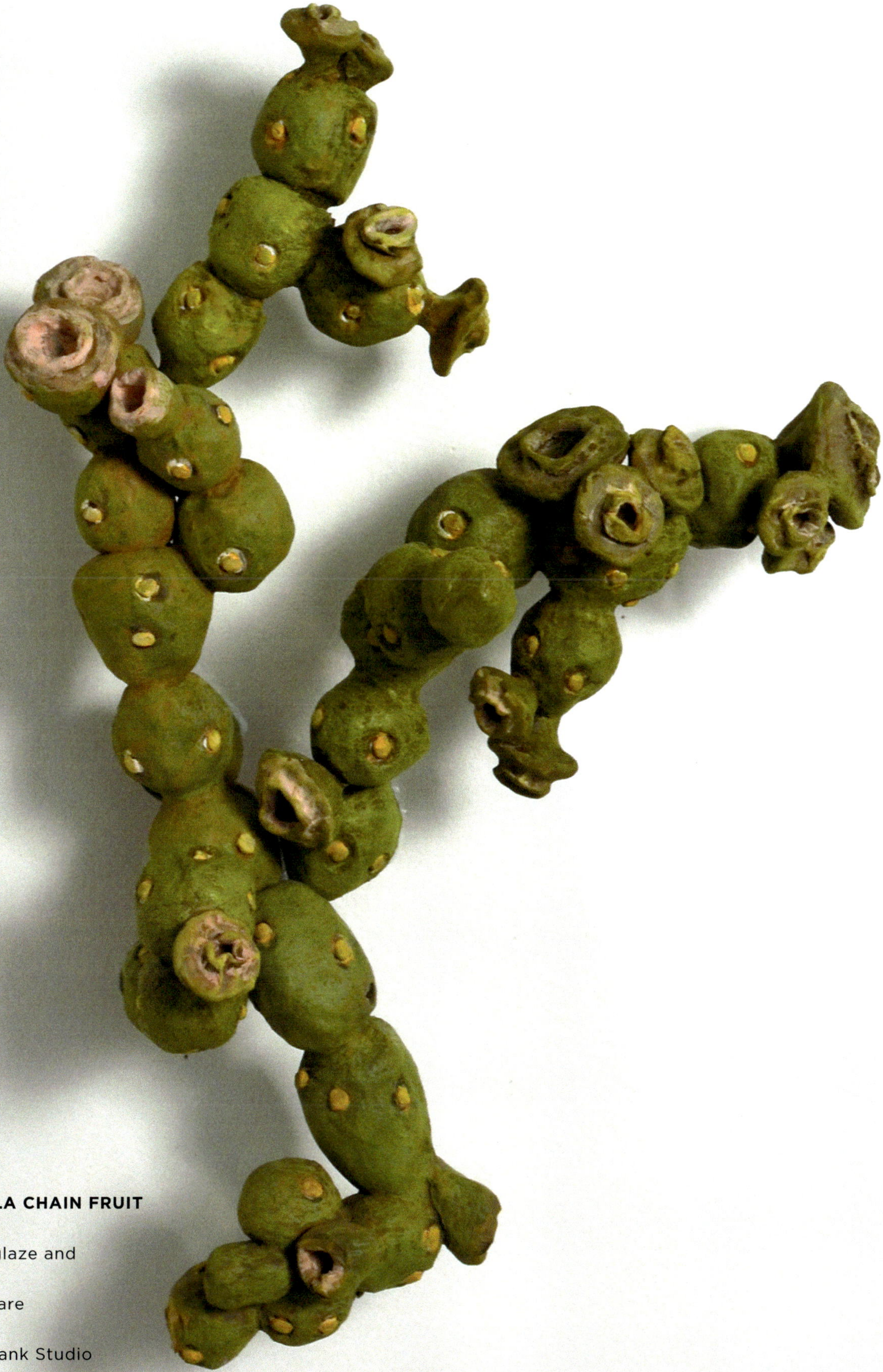

THRESHOLDS, CHOLLA CHAIN FRUIT CACTUS II, 2018
Ceramic cactus with glaze and colored pencil,
Plexiglas, glue, hardware
24 x 18 x 4 inches
Courtesy of Rachel Frank Studio

RACHEL FRANK

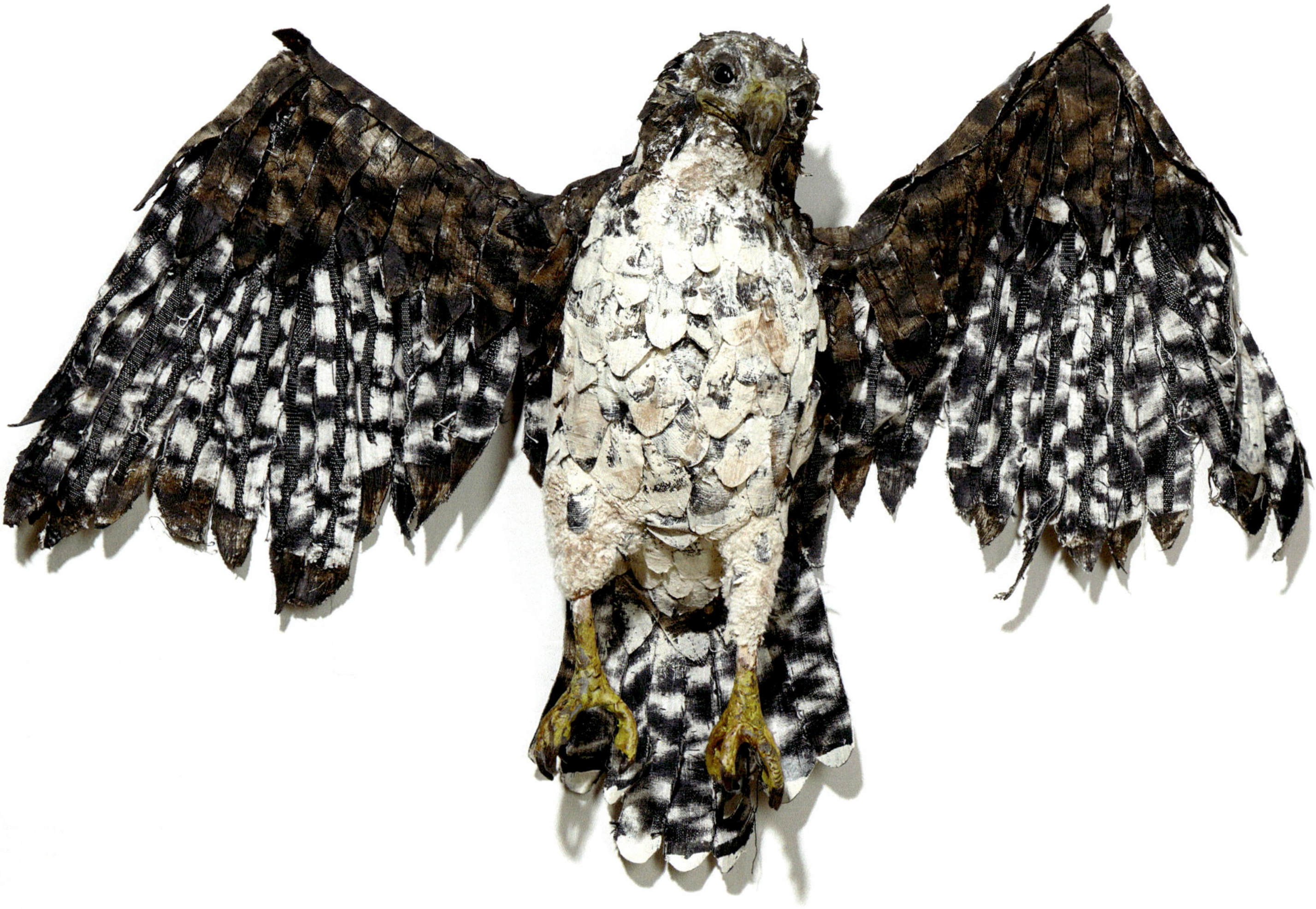

AUGURING HAWK, 2017
Fabric, thread, acrylic paint, wood, and other materials
22 x 36 x 9 ½ inches
Courtesy of Rachel Frank Studio

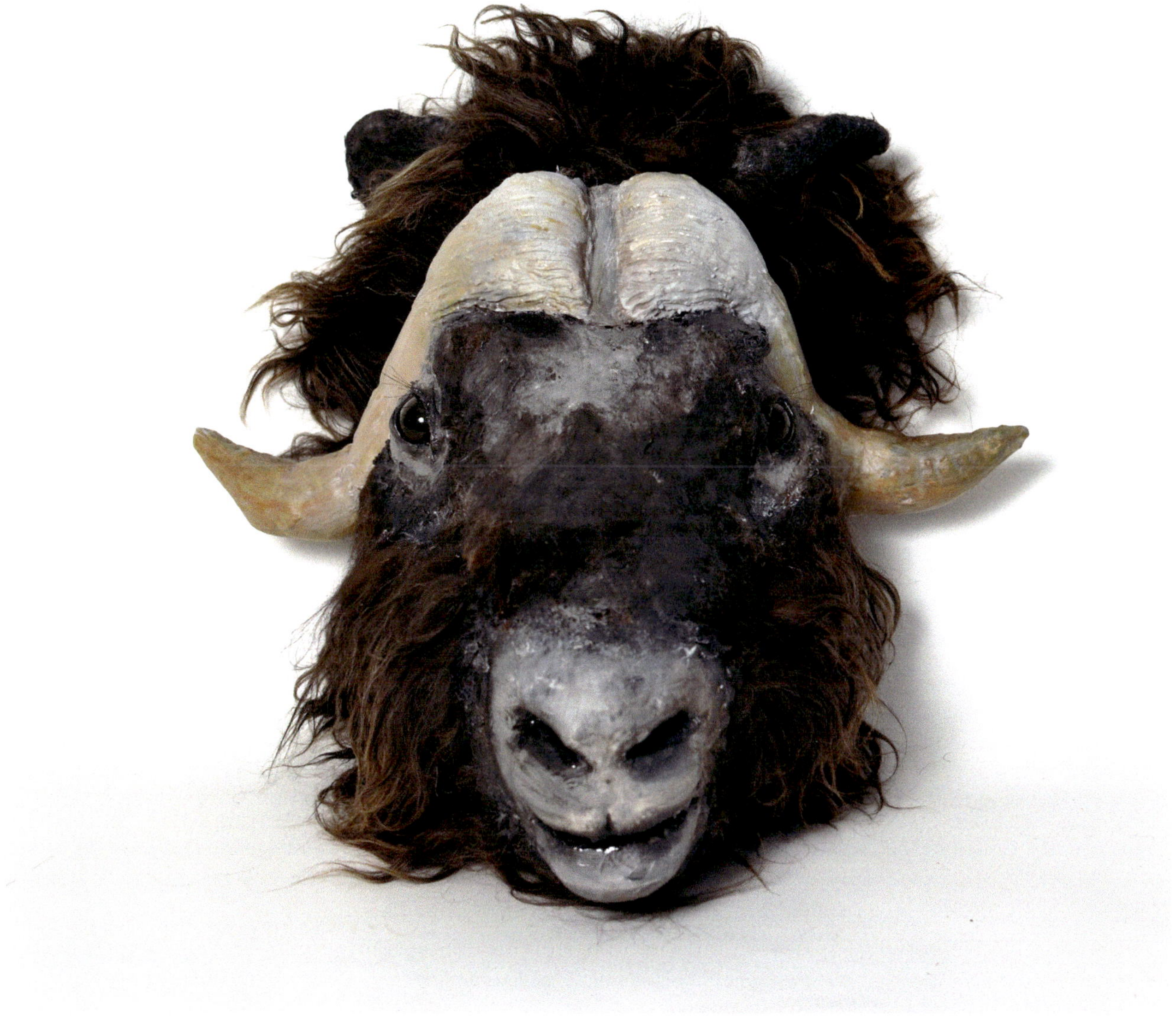

MUSK OX, 2016
Fabric, thread, wool, acrylic paint, and other materials
22 x 36 x 9 ½ inches
Courtesy of Rachel Frank Studio

RACHEL FRANK

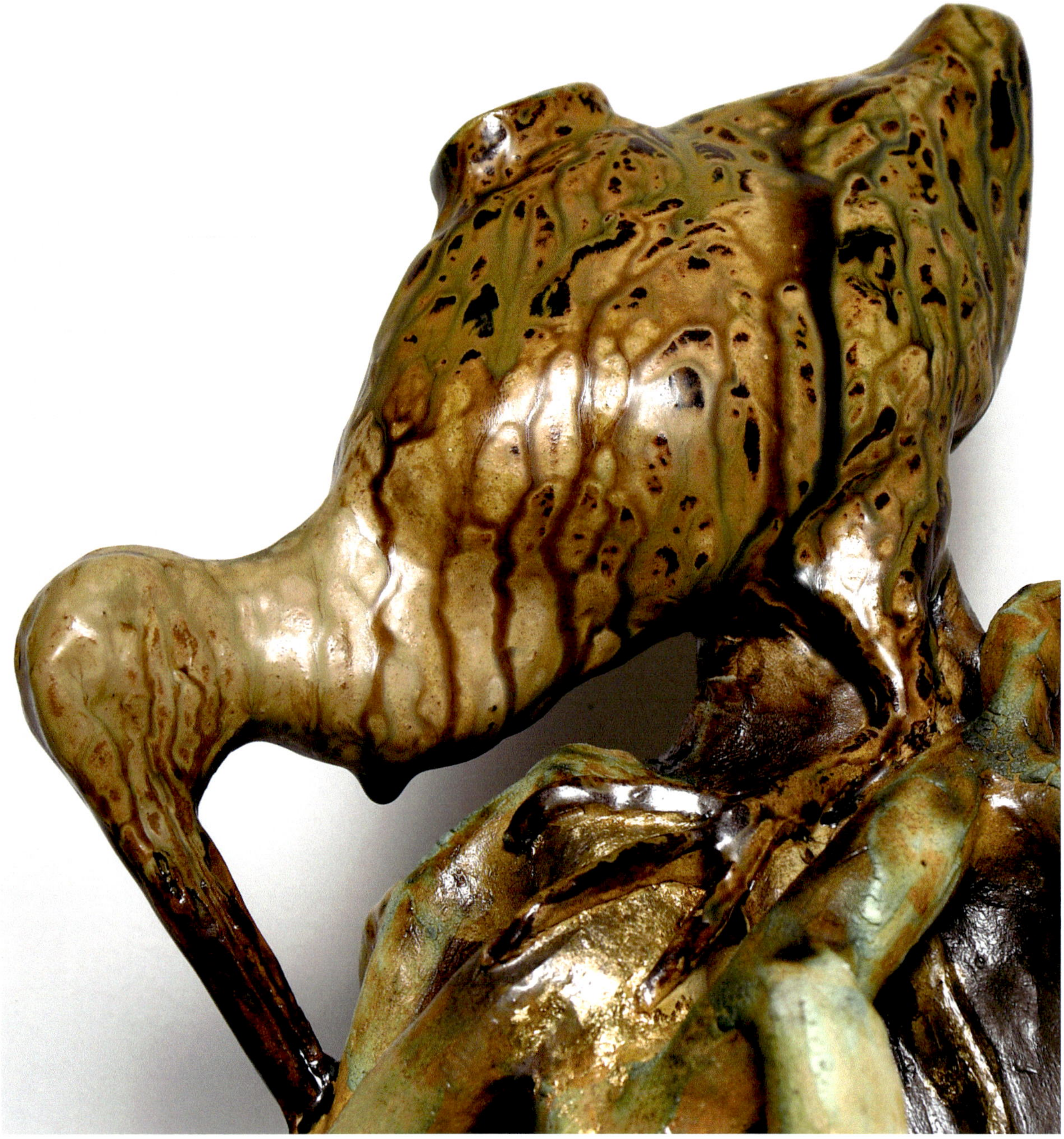

SENTINEL OFFERING LEKYTHOS:
CLAPPER RAIL AND MANGROVE, 2021
Stoneware ceramic and glazes
36 x 12 ½ x 14 inches
Courtesy of Rachel Frank Studio

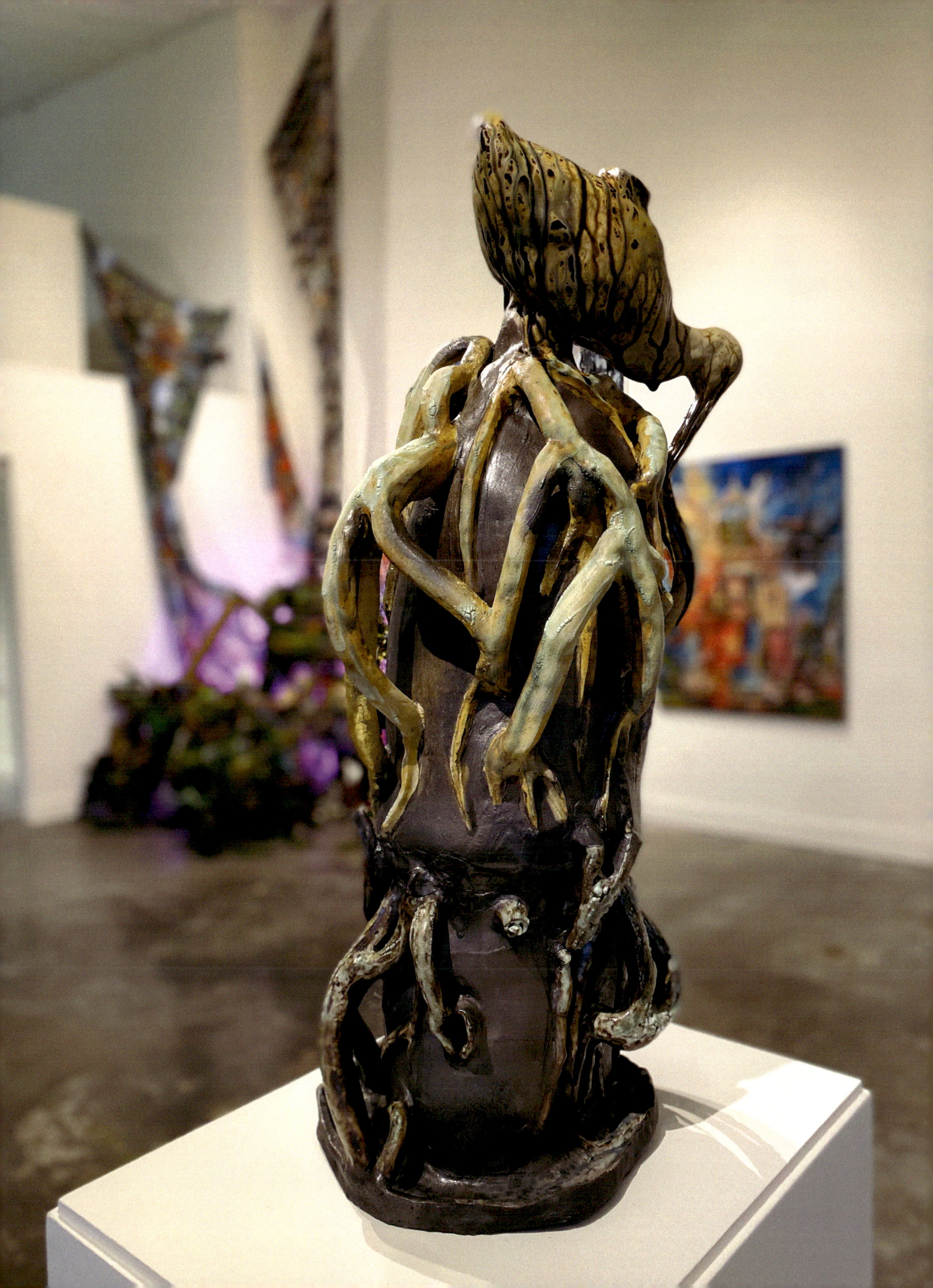

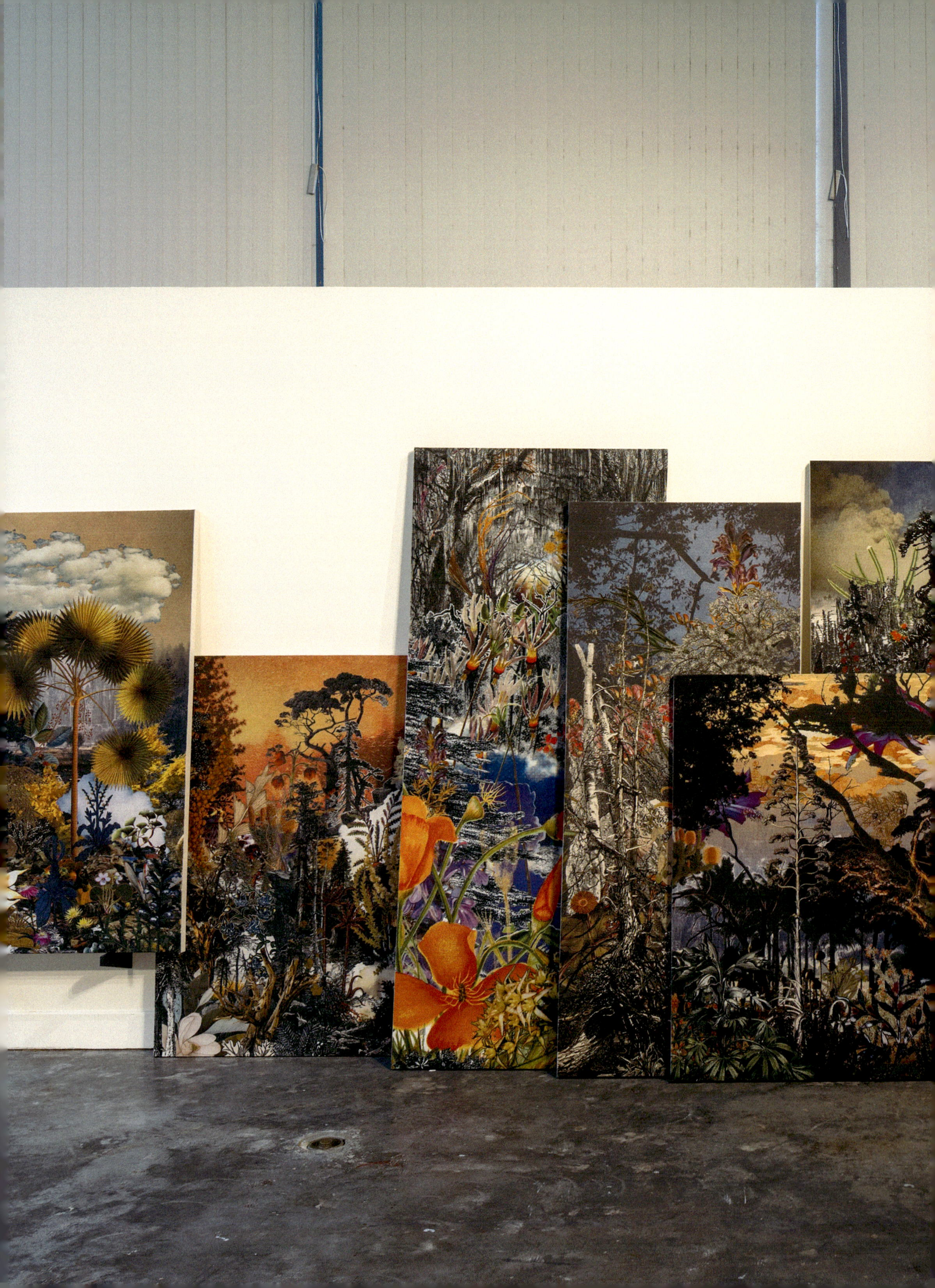

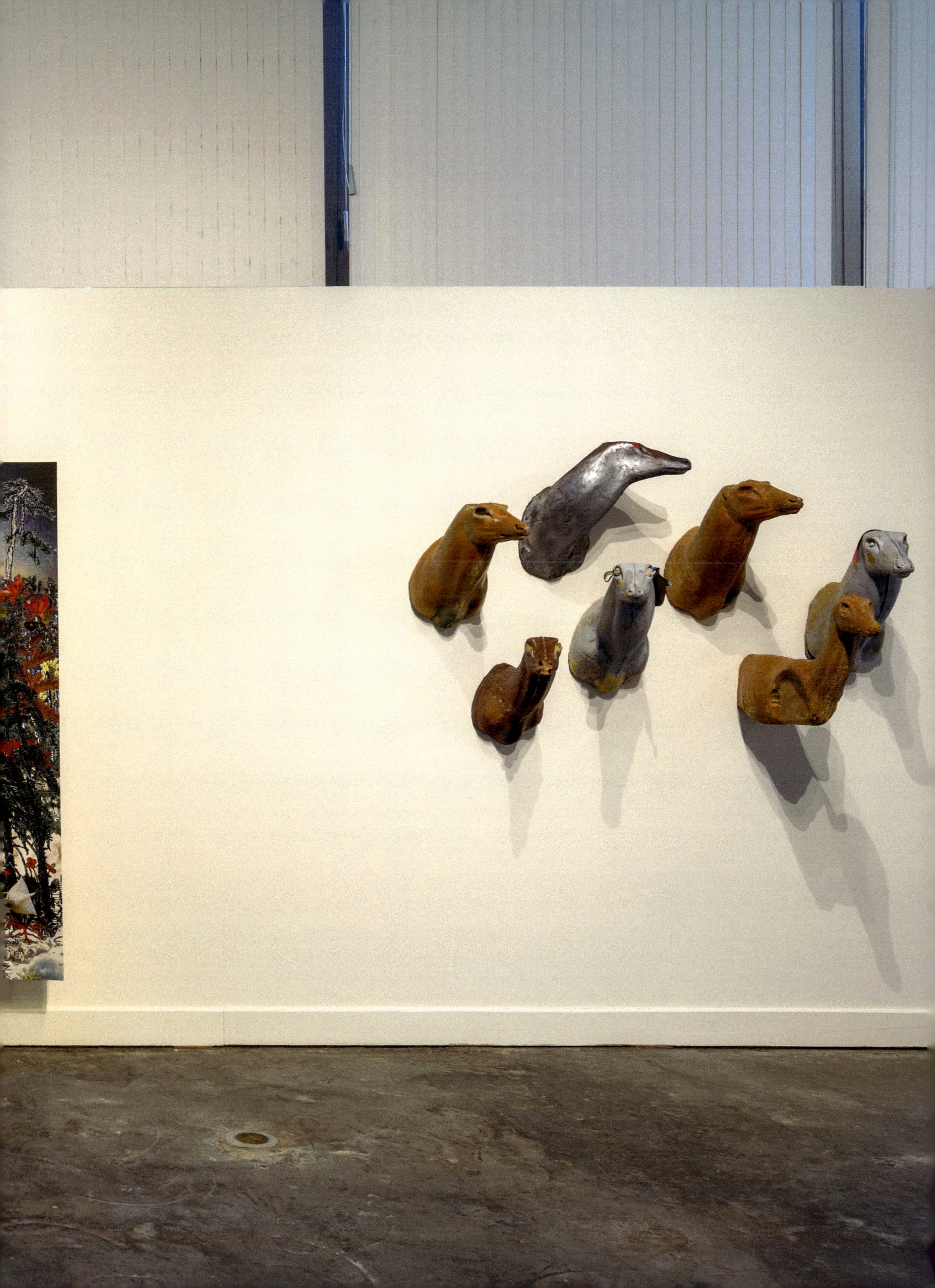

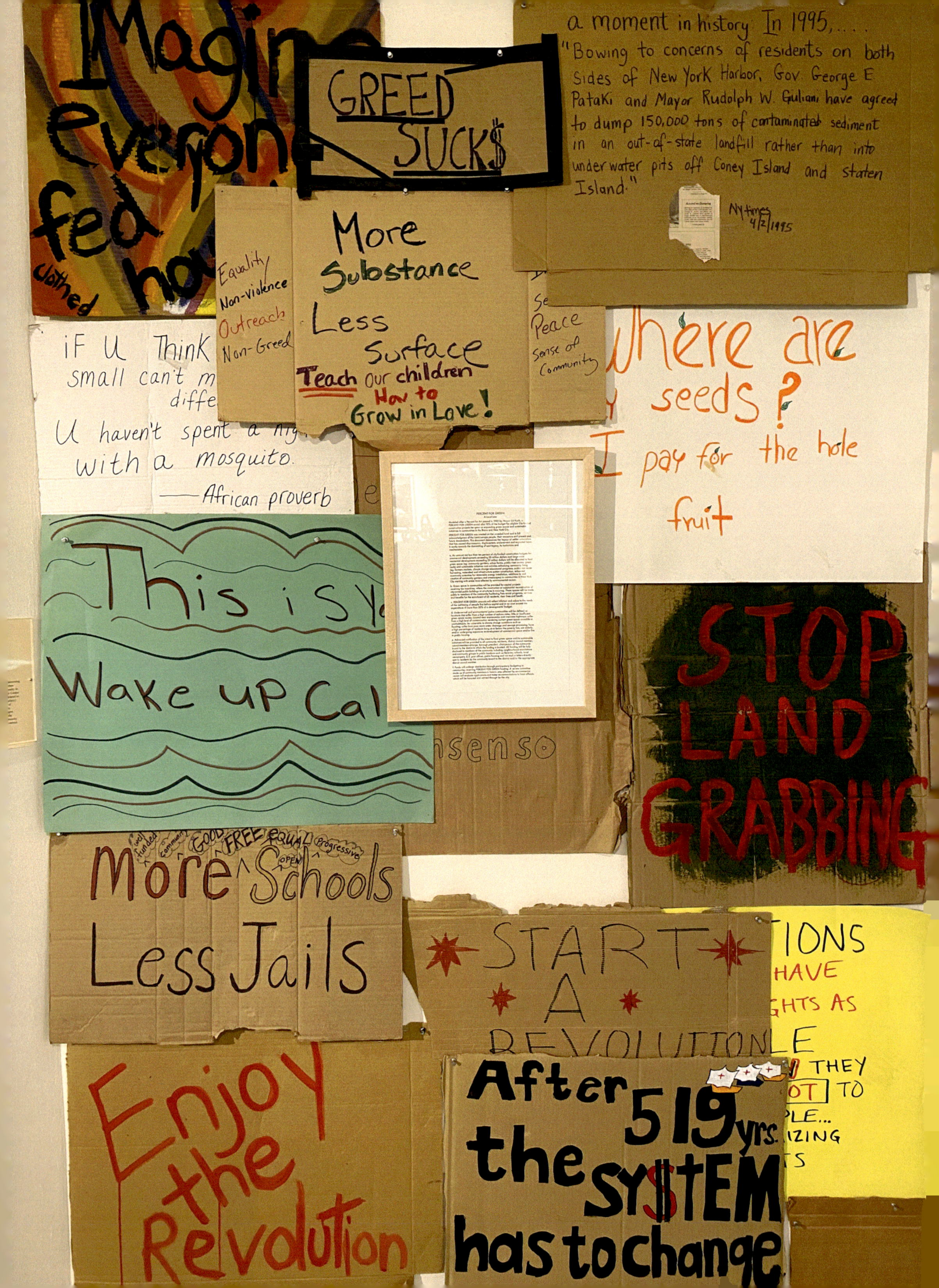
Imagin
everyon
fed
clothed
GREED SUCK$
a moment in history In 1995, ...
"Bowing to concerns of residents on both sides of New York Harbor, Gov. George E Pataki and Mayor Rudolph W. Guliani have agreed to dump 150,000 tons of contaminated sediment in an out-of-state landfill rather than into underwater pits off Coney Island and Staten Island."
NY times 4/2/1995
Equality
Non-violence
Outreach
Non-Greed
More Substance
Less Surface
Teach our children How to Grow in Love!
Peace
Sense of Community
if U Think
small can't m
diffe
U haven't spent a nig
with a mosquito.
—African proverb
where are
seeds?
I pay for the hole
fruit
This is Y
Wake up Cal
nsenso
STOP
LAND
GRABBING
funded
GOOD
FREE
EQUAL
Progressive
OPEN
More Schools
Less Jails
START
A
IONS
HAVE
GHTS AS
LE
THEY
OT TO
PLE...
IZING
Enjoy
the
Revolution
After 519 yrs.
the SY$TEM
has to change

PERCENT FOR GREEN BILL, 2014
Letterpress print on French paper, hung on protest ephemera (ink on cardboard, various dimensions); framed bill 14 x 17 inches, ephemera, dimensions variable
Courtesy of the artist

NOT PICTURED

SURGE, 2019
Single channel video
Courtesy of the artist

ALICIA GRULLON

Percent for Green Bill grew out of a participatory project that addresses the impact of environmental racism and climate change in the urban setting. Grullon worked with youth from the Bronx's Hunts Point community through a commission of The Point CDC and Lincoln Center Resiliency Project, which explored the resiliency of a community facing problems that come from climate change. The video *Surge* utilizes a dramatic movie preview format with Bronx imagery of cityscapes and aerial views along with a voiceover of a poem by Ayleah Evans, a teenager who participated in the program. The protest signs in the video speak to the urgency and cyclical nature of climate change issues faced in the Bronx and in other urban communities. The voices of many community members embodied here seek a shift in approaches that remedy environmental crises as well as solutions that can be applied locally and around the globe.

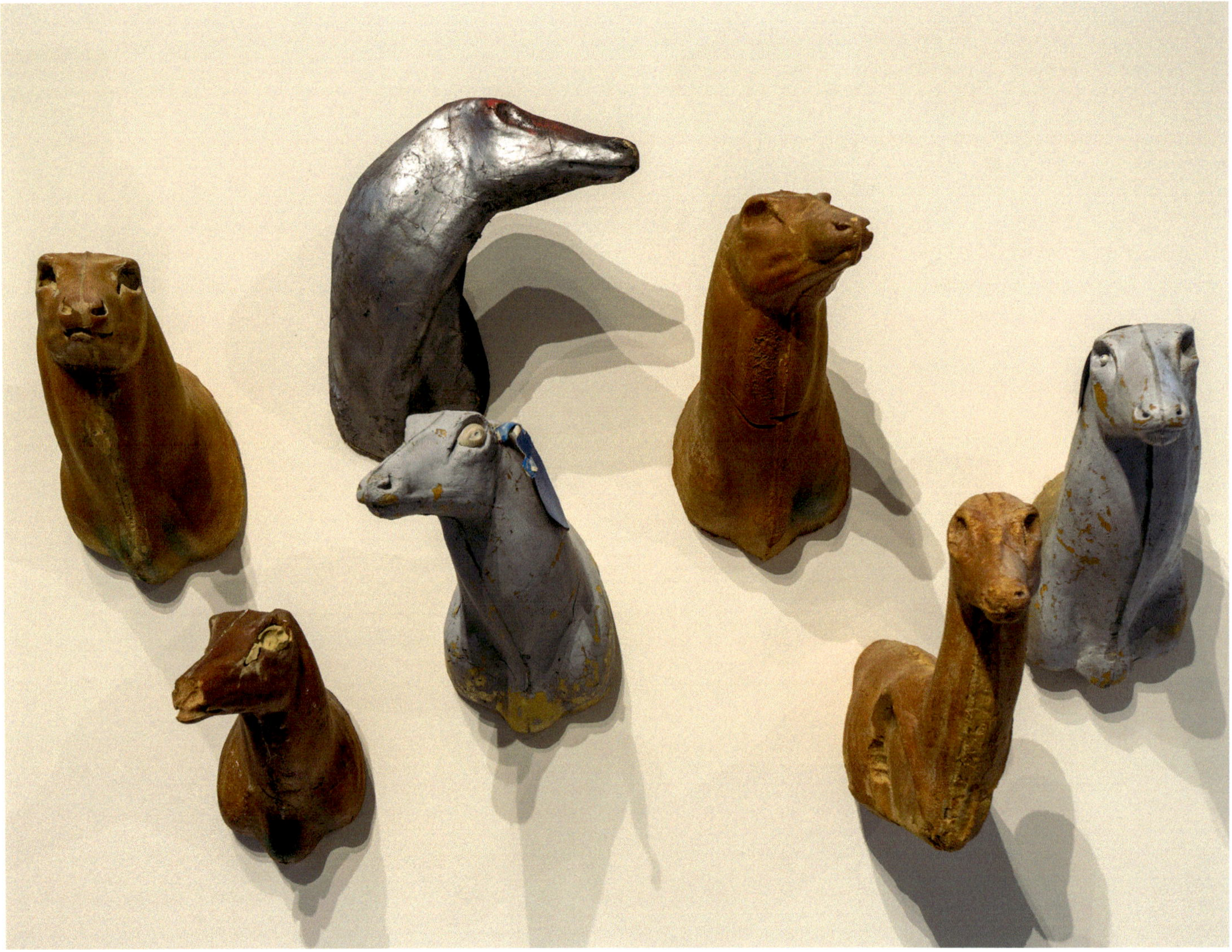

SEVEN CREATURES, 2017
Unique foam, mixed media
Dimensions variable
(AJH.16766)
Courtesy of the artist and Marianne Boesky Gallery,
New York and Aspen

ALLISON JANAE HAMILTON

Heads bursting from the Gallery wall hold facial expressions whose meanings are hard to discern. Do Hamilton's *Seven Creatures* accuse, implore, or threaten? To create this distinctive work, the artist used the same foam molds employed by taxidermists but she embellishes or disrupts the natural form of each animal in some way. Shown without their usual fur and color, these creatures become fantastical versions of common animals with a strangely "dinosaur-esque" quality, but they also appear skinned, denuded, and vulnerable. The title *Seven Creatures* refers to the White-tailed deer, native to Hamilton's home in North Florida. Taxidermy is commonly displayed across this country's southern states in households and communal spaces, but Hamilton's sculpted creatures that she made to appear taxidermied evoke an ecosystem that is tangible, endangered, and mythic. Speaking of her home state and its increasingly imperiled ecology, Hamilton says, "I grew up in the thick air, red soil and dark waterways of Florida, amongst mangroves, sawgrass rivers, and cypress swamps. Swamps have been disparagingly depicted by Virgil and Dante as lands of the 'unfortunate dead,' and more recently in our political landscape as a metaphor for corruption that must be 'drained.' However, marshes and wetlands are intrinsic protections against hurricanes and storm floods, and teem with biodiversity operating in continuous symbiosis."

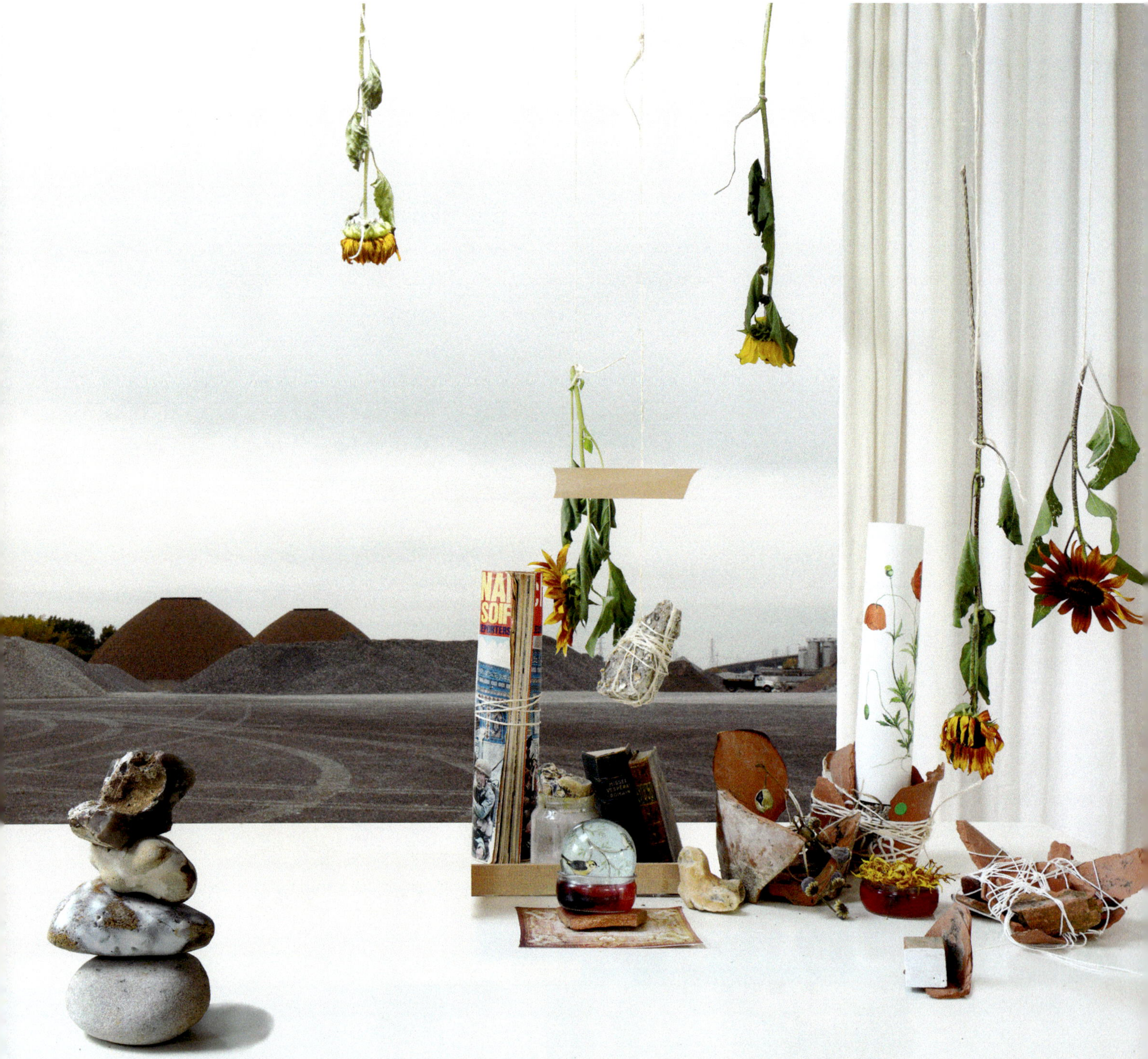

ASH FLOWER, 2017
Chromogenic dye coupler prints
30 x 30 inches
Courtesy of Robert Mann Gallery

MARY MATTINGLY

Mattingly's photographs allude to complex ideas of extinction. She balances the barren background of her landscapes with an assemblage of objects, flora, and fauna in the photograph's foreground to create both somber still lifes and poetic messages about our world. The motifs she chooses represent the toxic results of mining and chemical use. Mattingly specifically explores the contradiction of the photographic process and its impact on the environment, while referencing the supply chain of materials needed to produce a photograph. The play of shadows cast by her still lifes suggest fleeting memories of the natural world that seen through a window look barren, like the surface of another planet — one with no ability to promote life. On close observation, Mattingly echoes the nature of mining by using pilings of rocks and minerals. The flowers in her scenes are bound, constricted, and hung from above, reading as captive nature. These profound contradictions and entanglements create multiple layers of content in the photographs embedded within her innovative use of still life.

MARY MATTINGLY

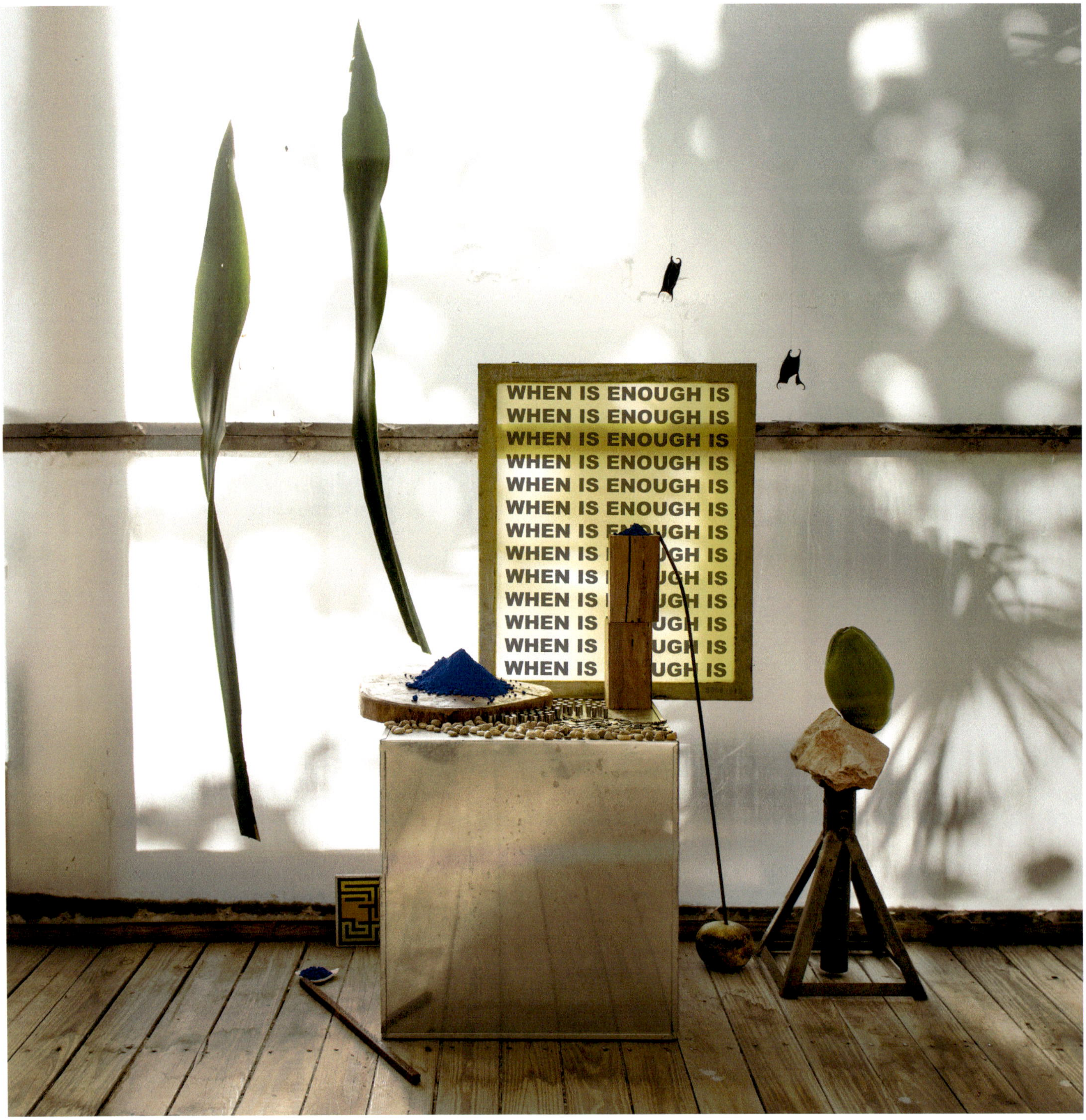

ENOUGH IS WHEN ENOUGH, 2018
Chromogenic dye coupler prints
30 x 30 inches
Courtesy of Robert Mann Gallery

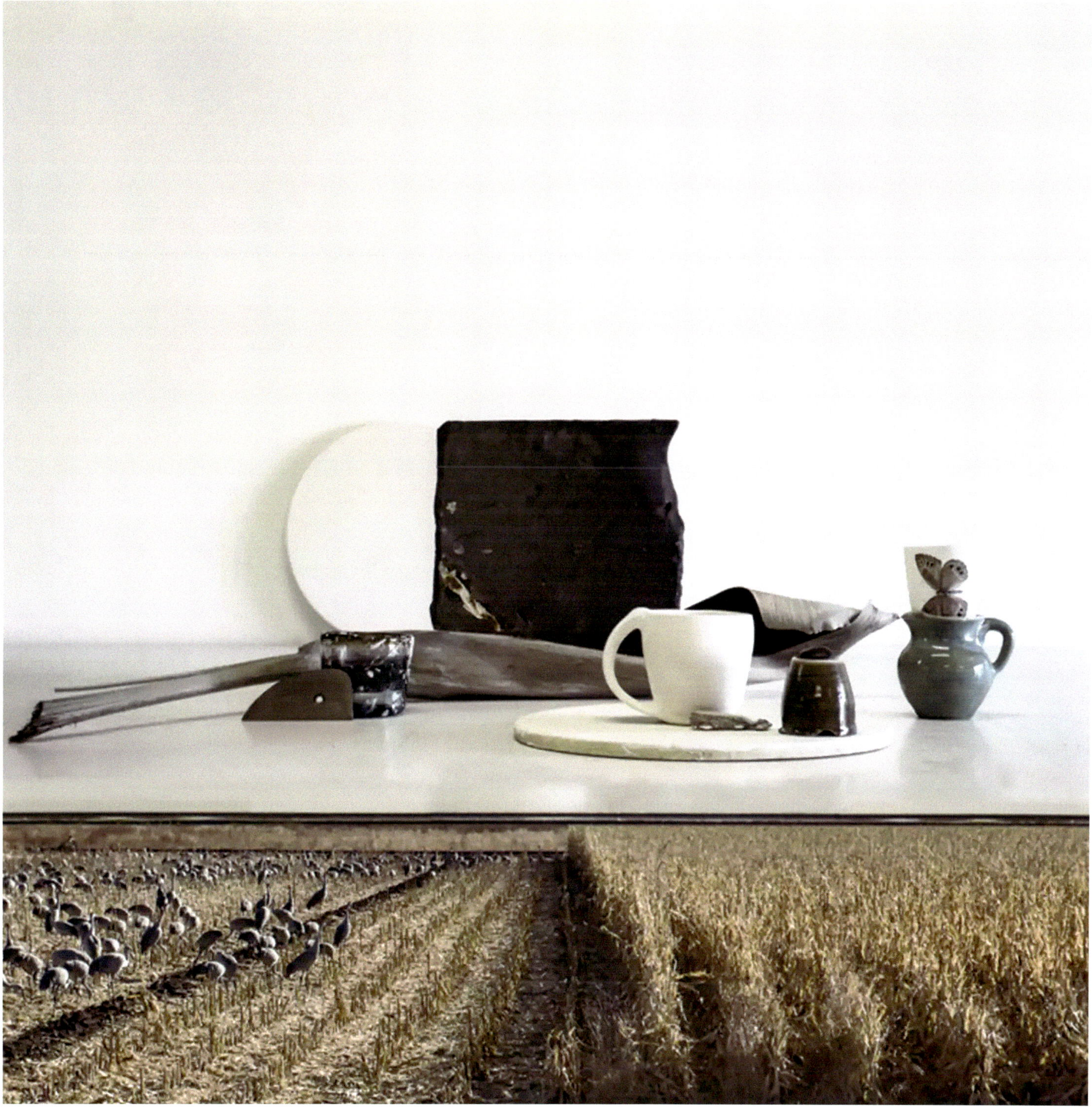

HELD MY BREATH AND DIDN'T HEAR ANYTHING, 2018
Chromogenic dye coupler prints
30 x 30 inches
Courtesy of Robert Mann Gallery

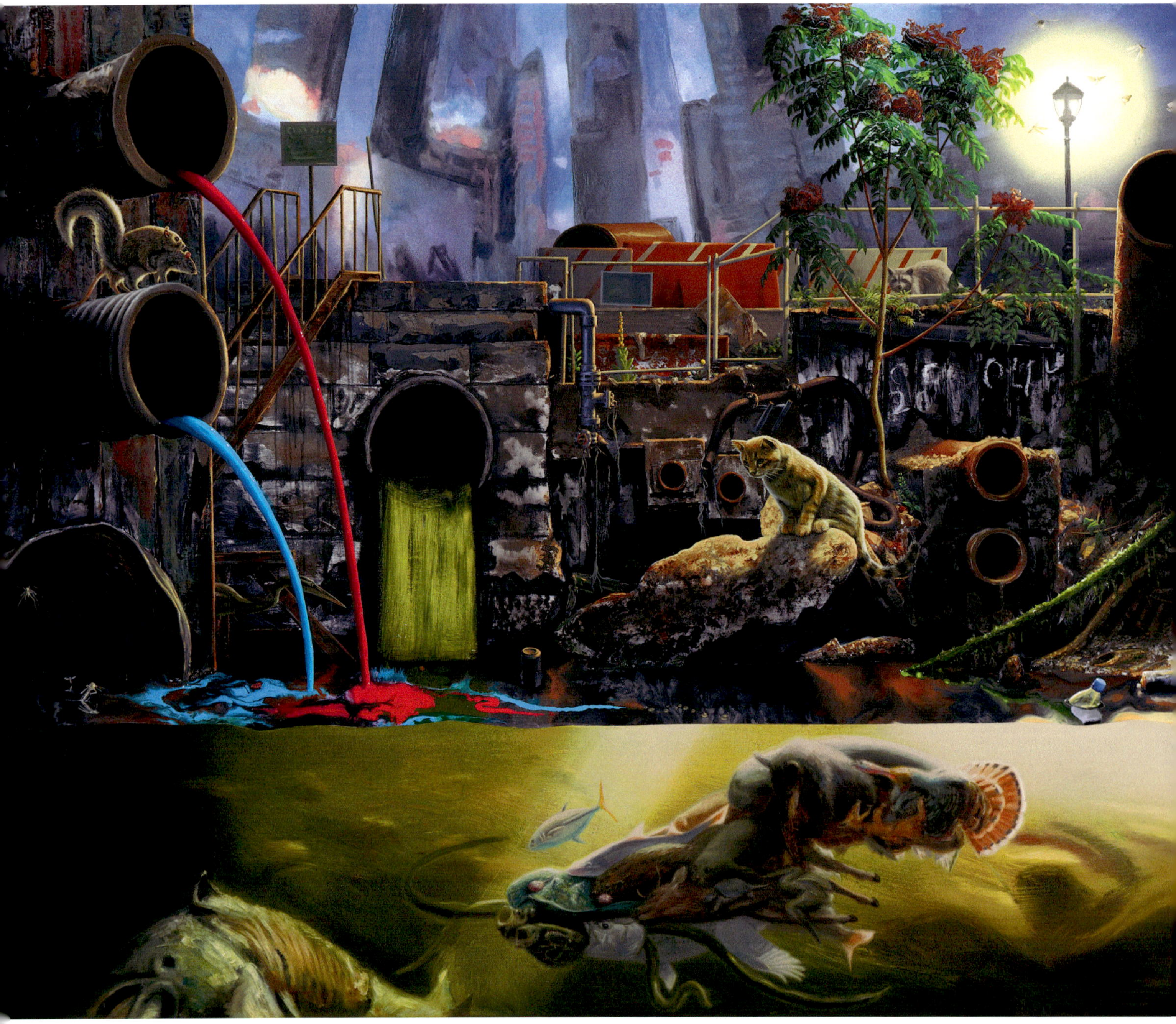

GOWANUS, 2013
Oil on wood
72 x 90 inches
Courtesy of Sperone Westwater

ALEXIS ROCKMAN

Rockman presents a canvas at the scale of a traditional history painting, encouraging the viewer to see a cycle of history in the form of Brooklyn's Gowanus Canal, infamous for its pollution. He arranges his composition scrupulously, bisecting the painting horizontally into above and below water scenes. Perched on a rock, a watchful cat observes the water below, where a disturbing beast lurks in the chemical soup of the canal. Rockman has described this animal as a composite creature made from animals that used to live in the canal. "It's sort of the ghost of the Gowanus Canal in the shape of a catfish. . . . I became obsessed with composite creatures and this whole tradition in Indian miniature paintings of a big creature made out of little creatures. Here, the animal is comprised of a diamondback terrapin, a black bear, a manatee, an American turkey, a red fox, a harbor seal, and so on. The painting was really inspired by a dolphin that swam into the canal and proceeded to die. I was sad and then thought how many things have died there over the years? Alexis Rockman also links the toxicity of the canal and the chemicals found in many art supplies — the drains that pour into his canal are filled with vibrant red, yellow, and blue primary colors, suggesting poisonous pigments added to the artist's palette. He suggests that the act of creation comes with a price.

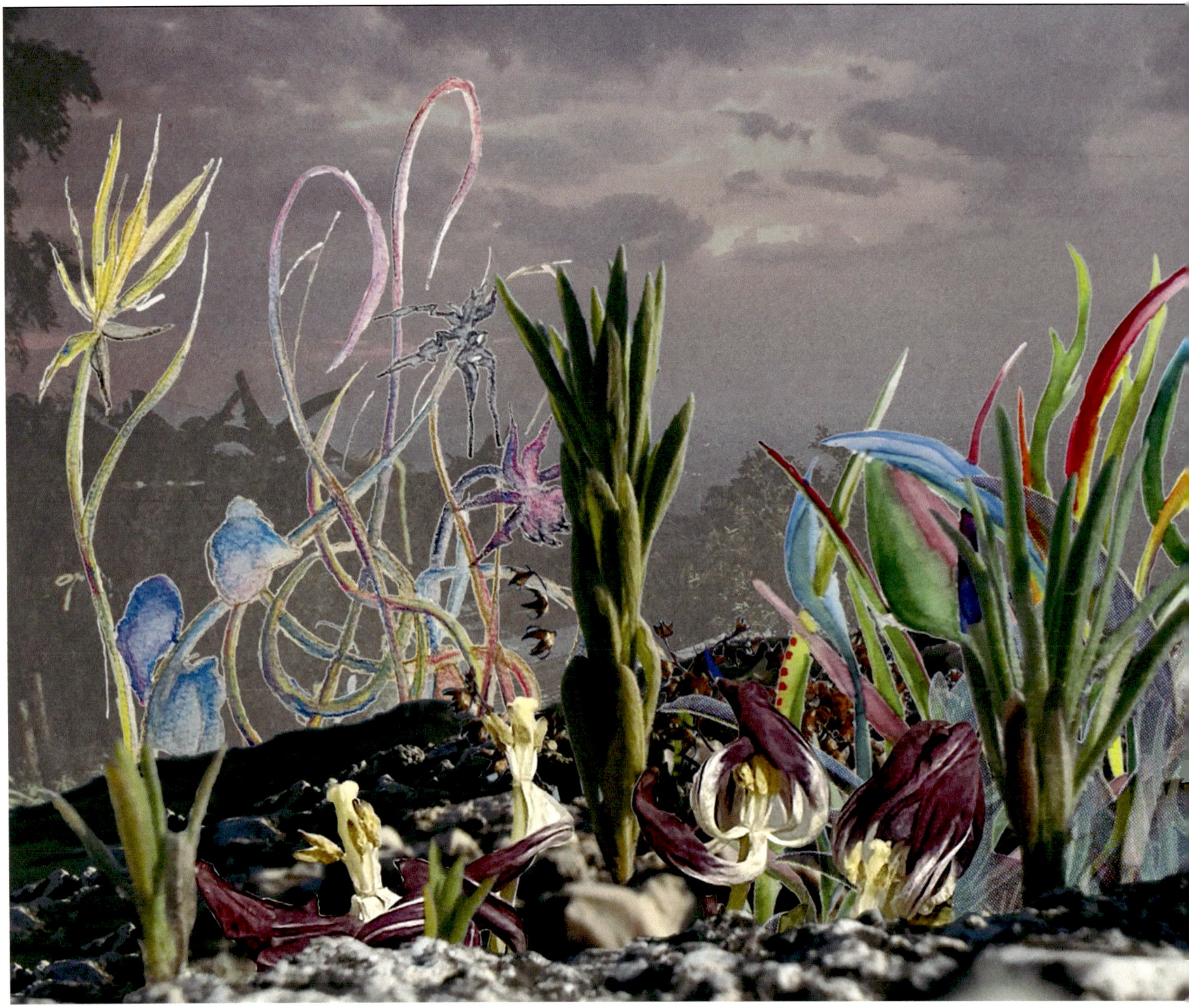

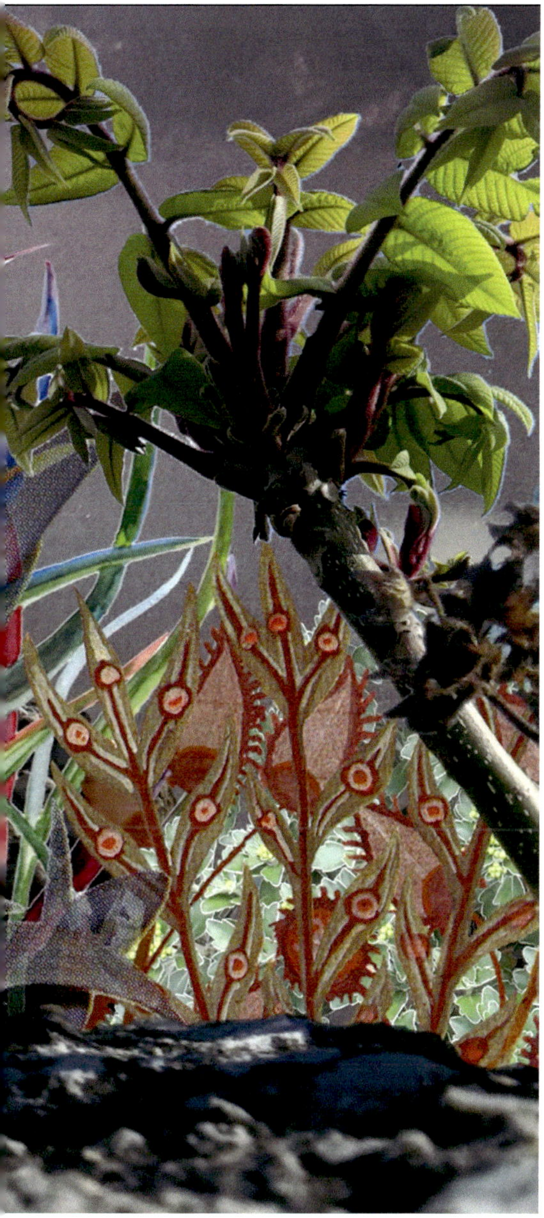

UNRELENTING, 2020
Video mp4
5 minutes
Courtesy of the artist

FRANCESCO SIMETI

Simeti created this video while sheltering in place in New York City, during the Covid-19 pandemic lockdown. Developing animations, he depicted the emergence of the natural world, just as humans retracted indoors. The time-lapse quality of the film shows flora and fauna emerging and receding in slow rhythmic patterns. Christopher Cerrone, a composer, collaborated with Simeti for this piece, and his music composition adds to its meditative and foreboding mood. The mournful quality of its imagery and sound captivates the viewer and begs the question of an unexpected and unquestionable disruption of the urban landscape as we know it.

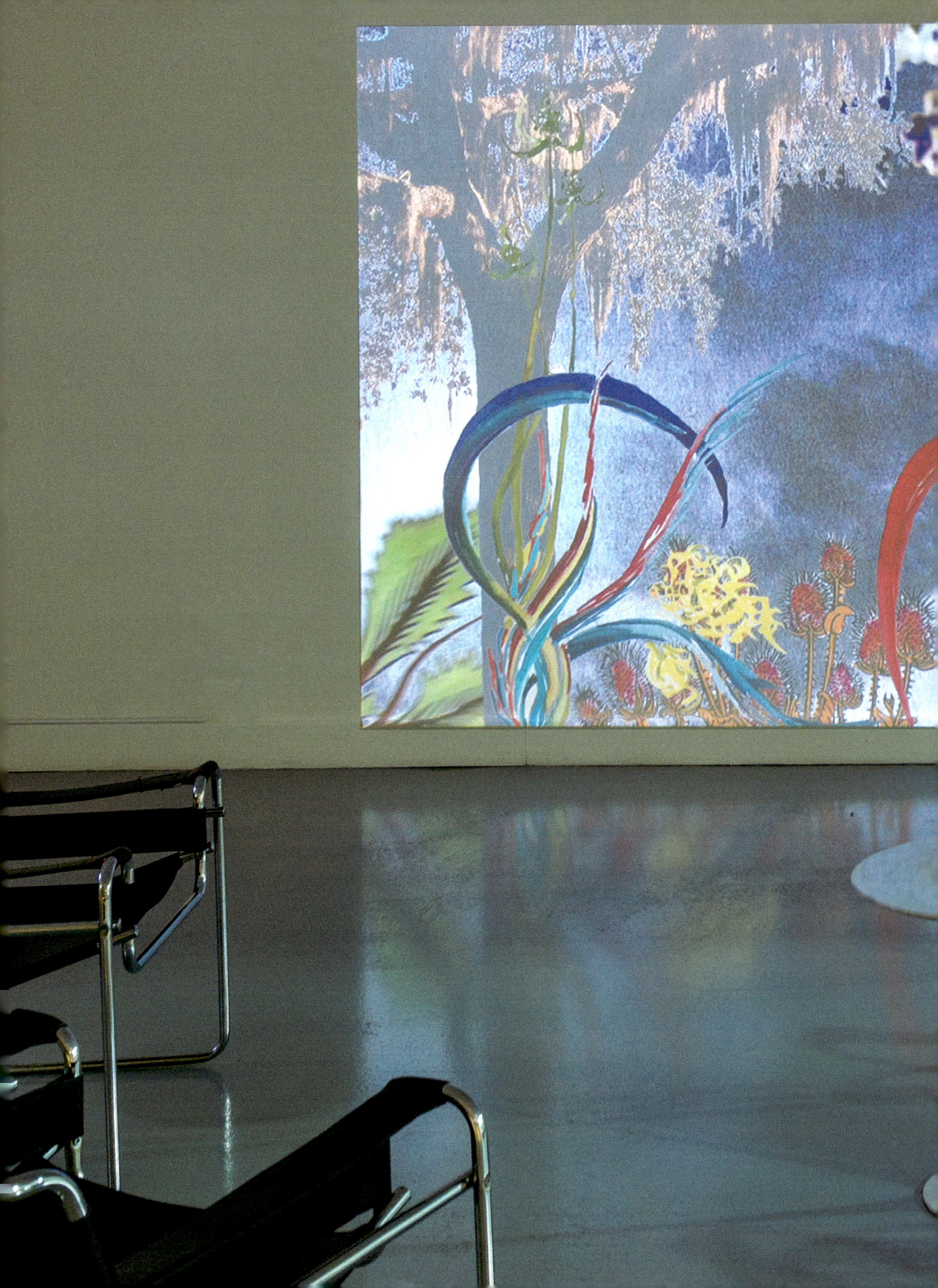

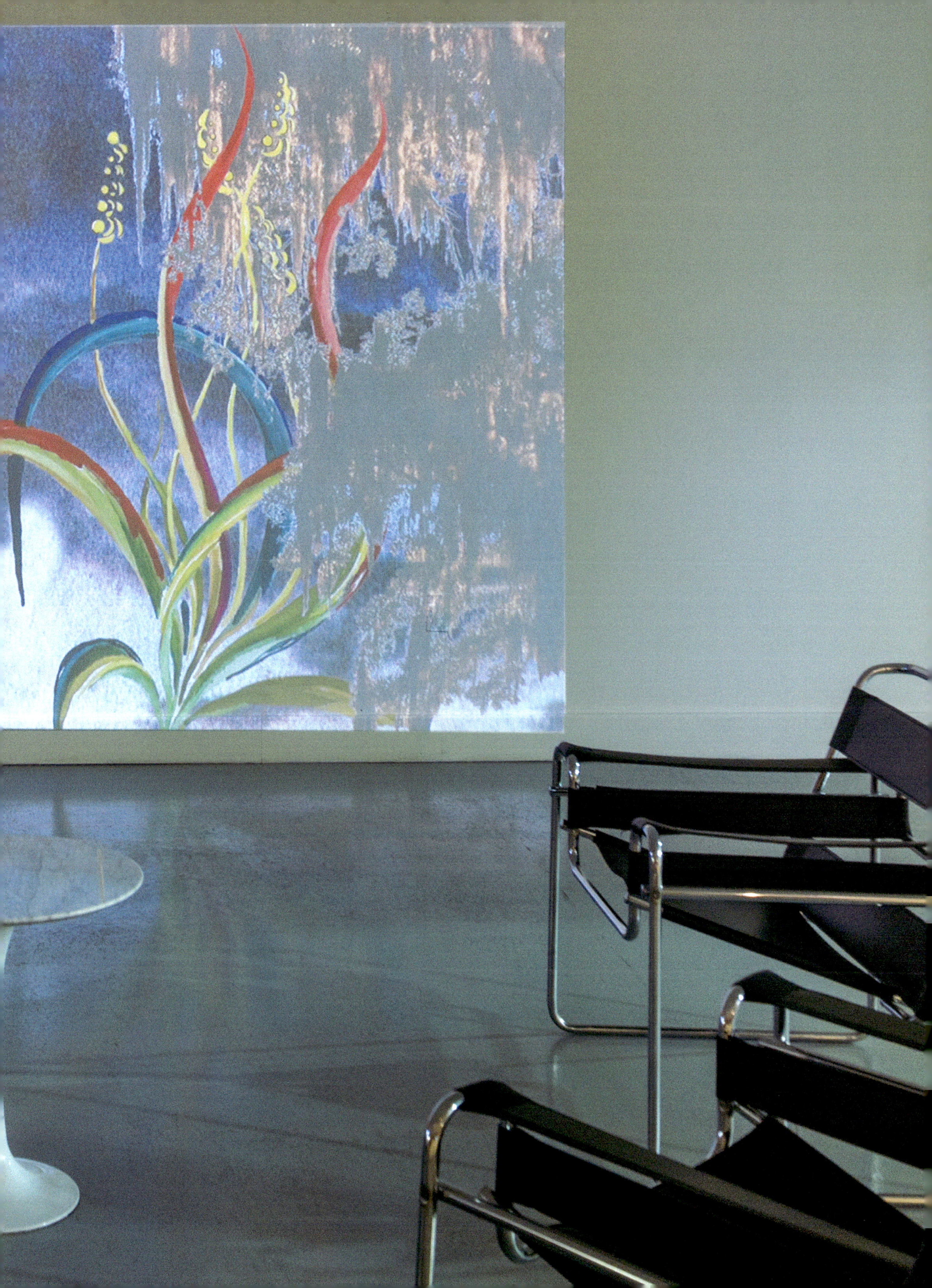

FRANCESCO SIMETI

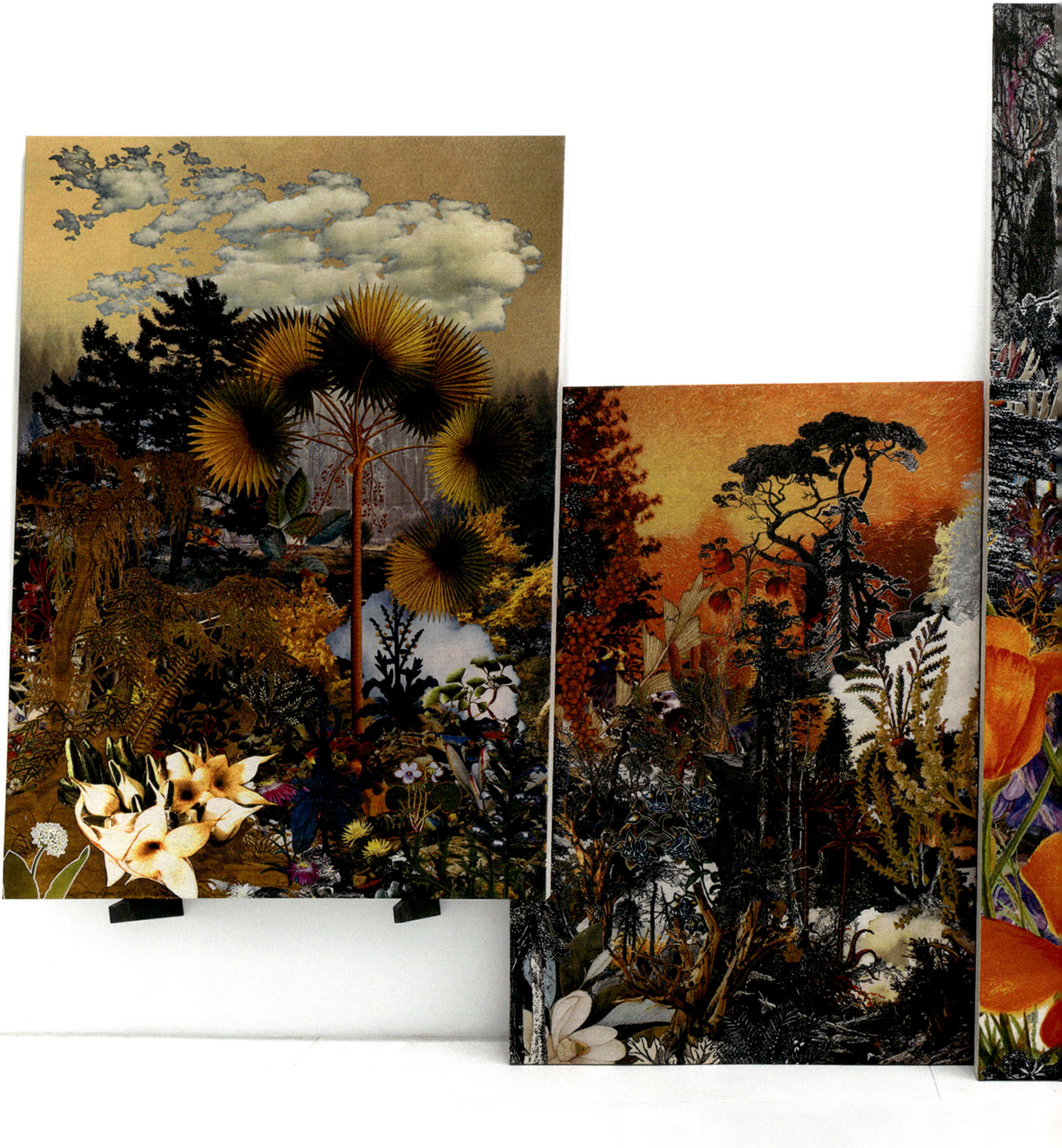

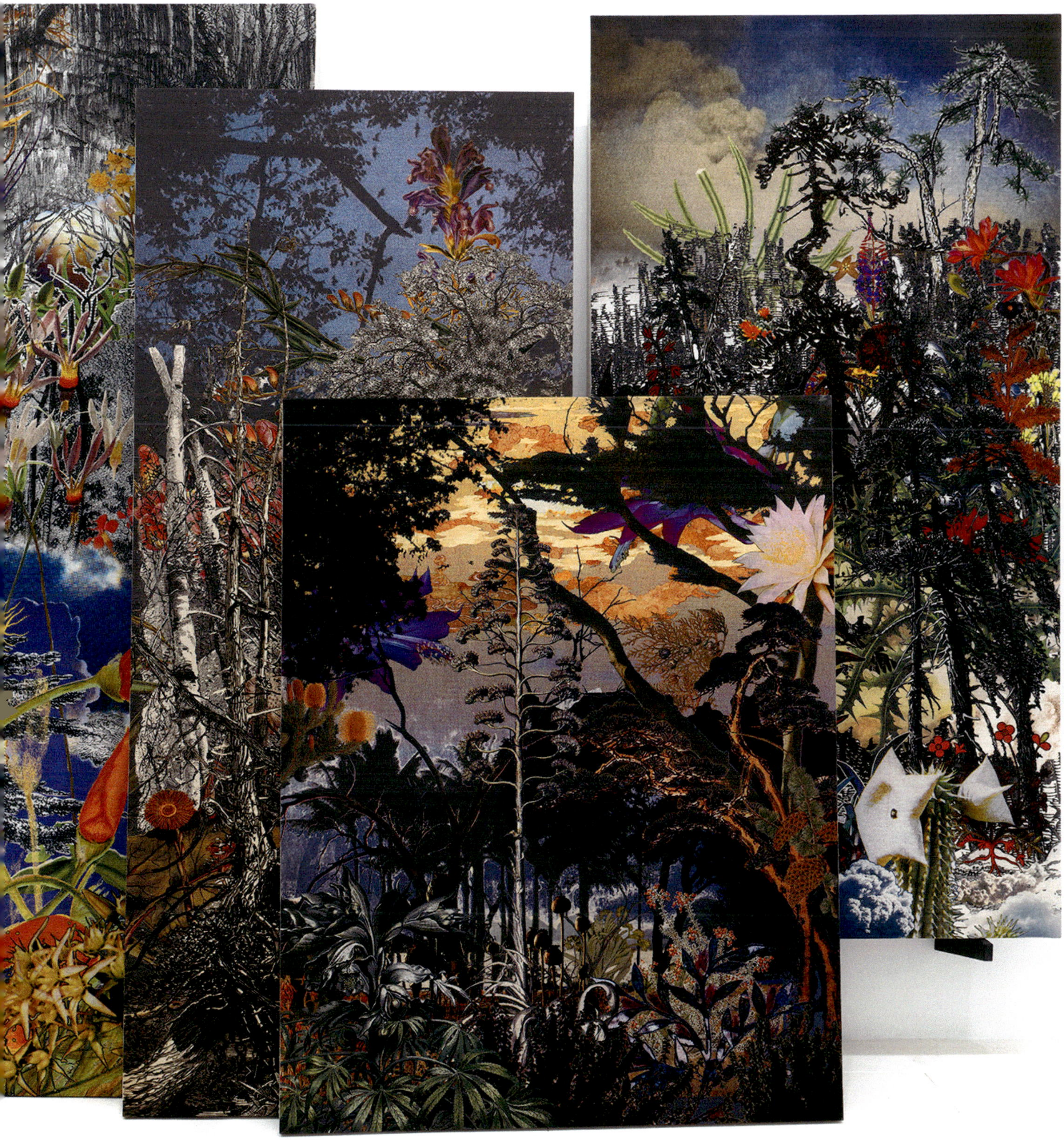

FRANCESCO SIMETI

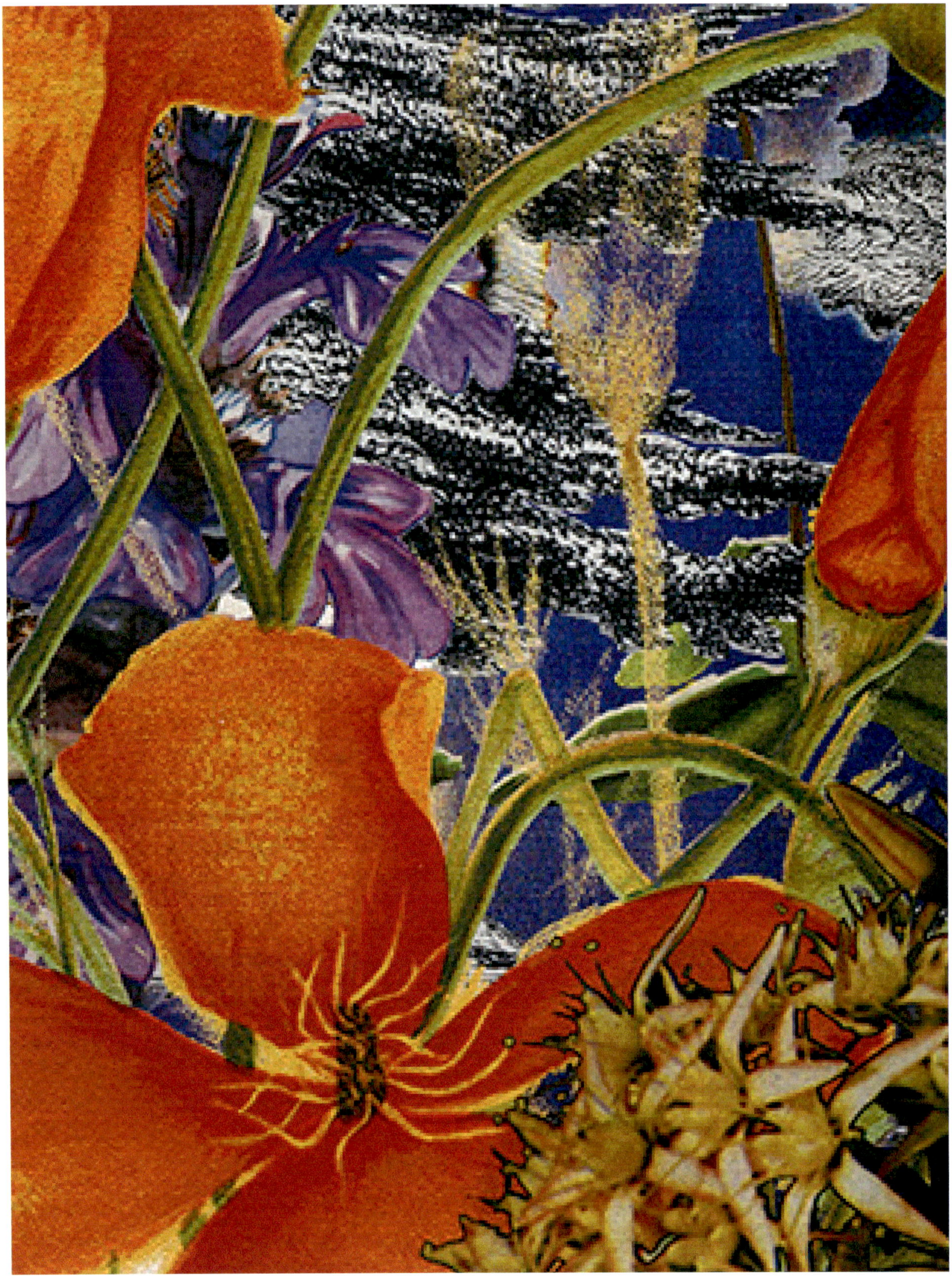

REFUGIUM (PHYLLOCACTUS FROM SUMMIT), 2019
Detail. Digital print on linen
78 x 40 inches
Courtesy of the artist

Refugium refers to a place to hide and be protected from dangers, specifically a place where non-humans can live without fear. In this digital print, Simeti contemplates the strains between humans and the natural world and examines how humans negatively impact the non-human world. He sourced imagery from newspapers, magazines, and botanical illustrations to create *Refugium*, a fragile landscape he filled with native and exotic flora. It is a somber environment without humans, desolate, yet a riotous landscape, where all living things thrive. This composition highlights the violent reality inflicted upon nature by humans and by the pressures of our economic structures.

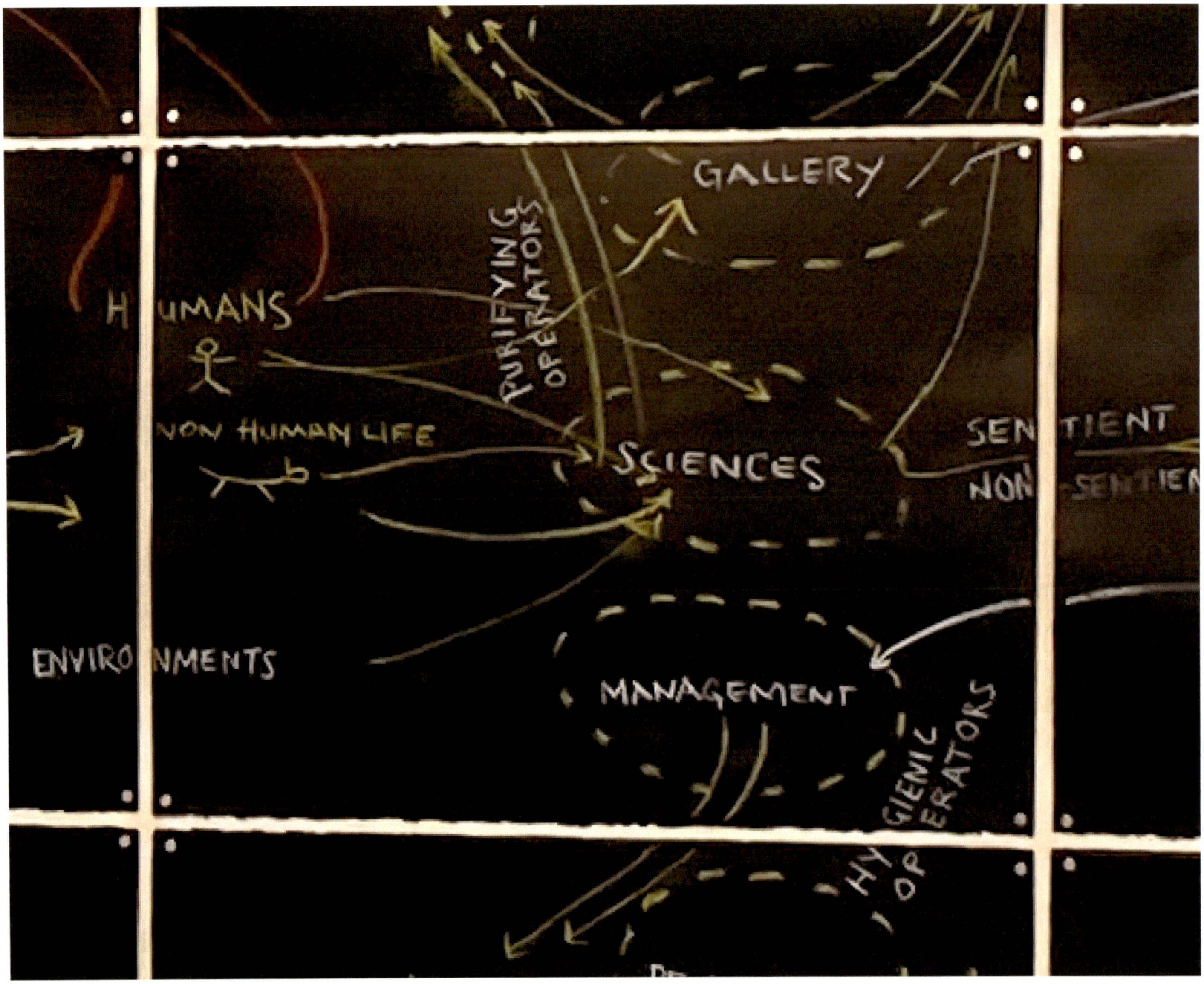

MORE THAN US THRIVING WHERE WE ARE, 2020
Dimensions variable
Courtesy of SPURSE,
Block Ecology, and Flourish Lab

THIS PAGE AND THE FOLLOWING SIX PAGES

SPURSE

SPURSE, Block Ecology and Flourish LAB in collaboration with Jack Henning, PhD and Renuka P. Sankaran, PhD, Department of Biological Sciences, Lehman College, City University of New York, Bryan Williams and Jezreel Deseo

The interdisciplinary collectives SPURSE, Block Ecology and Flourish LAB work with communities and institutions to develop strategies to center interdependency with nature. Their site-specific, research-based project, presented at both Lehman College and Wave Hill, explore untapped aspects of living in an urban environment, including potential food sources. As a collective, the goals of SPURSE state: "We believe that art about nature is not enough; if we are to engage our ecologies in a meaningful way, we need to authentically become of the places in which we live. For us, the solution lies in a far-reaching transformative place-based rethinking of social, ecological and artistic methods. The problem today is that our transience from place to place seriously exacerbates pressing social and ecological issues. . .. Our practice of change starts by collaborating with the thriving ecology that is right under our feet. In our dense urban areas, there are ecosystems that have the potential to generate new forms of community and resilience. We largely ignore these areas of spontaneous growth preferring our cultivated and controlled landscapes. Migrating with people, "wild" species have co-evolved with us for millennia and until our present generation, foraging was what linked communities together, connecting us deeply and profoundly to each other and our environment in an active manner." Here, SPURSE presents a large-scale drawing, originally presented at the Museum of Contemporary Art, that decenters the human experience and focuses on life other than human, while also presenting a living collection of plants, commonly called "weeds", which grow in our local urban environment. By elevating these plants to a gallery setting, we are forced to contemplate why some plant life is considered "undesirable."

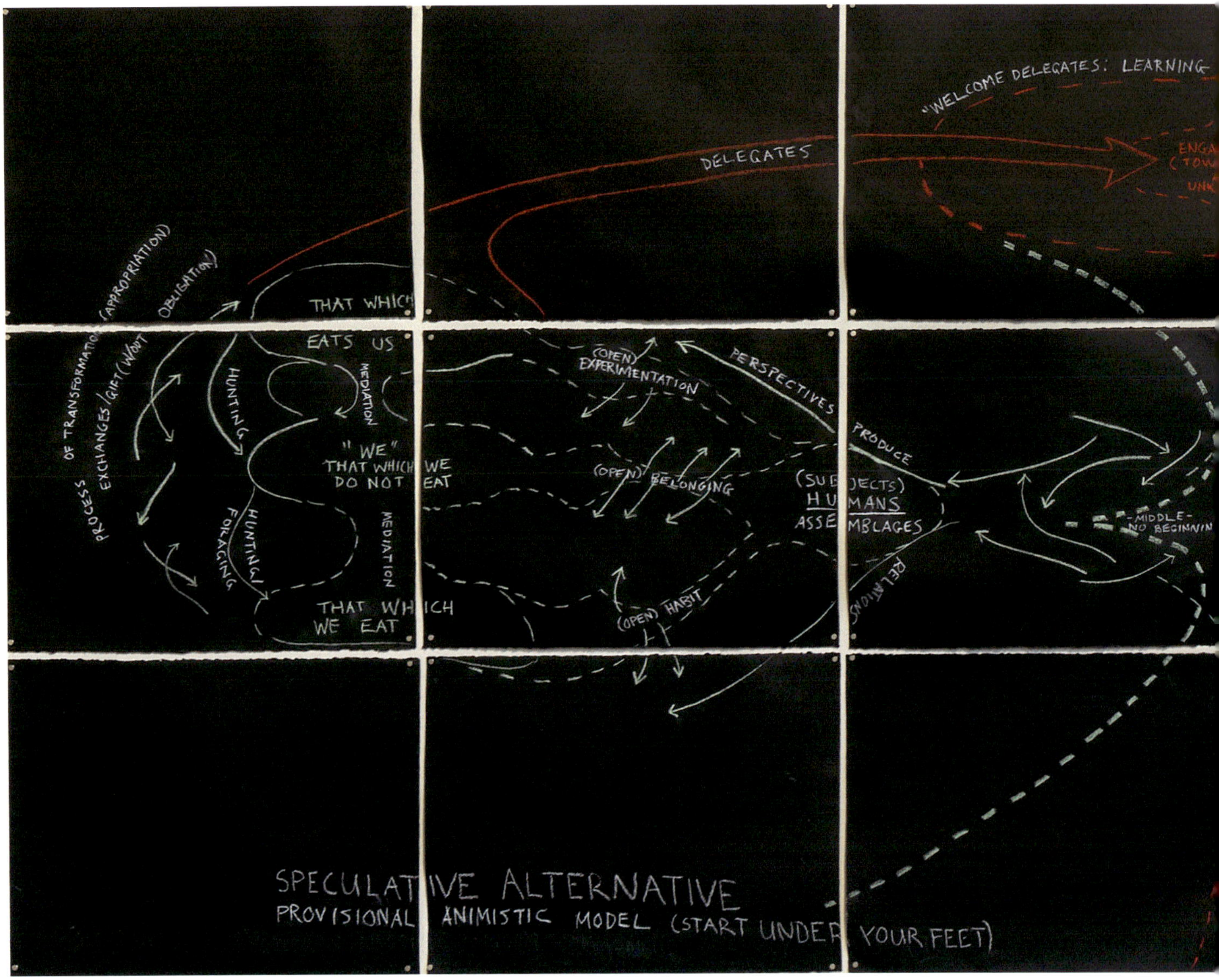
WELCOME DELEGATES: LEARNING
DELEGATES
THAT WHICH EATS US
(APPROPRIATION)
OBLIGATION)
PROCESS OF TRANSFORMATION
EXCHANGES/GIFT (W/OUT
HUNTING
MEDIATION
"WE"
THAT WHICH WE DO NOT EAT
(OPEN) EXPERIMENTATION
PERSPECTIVES
PRODUCE
(OPEN) BELONGING
(SUBJECTS)
HUMANS
ASSEMBLAGES
HUNTING/ FORAGING
MEDIATION
THAT WHICH WE EAT
(OPEN) HABIT
RELATIONS
-MIDDLE-
SPECULATIVE ALTERNATIVE
PROVISIONAL ANIMISTIC MODEL (START UNDER YOUR FEET)

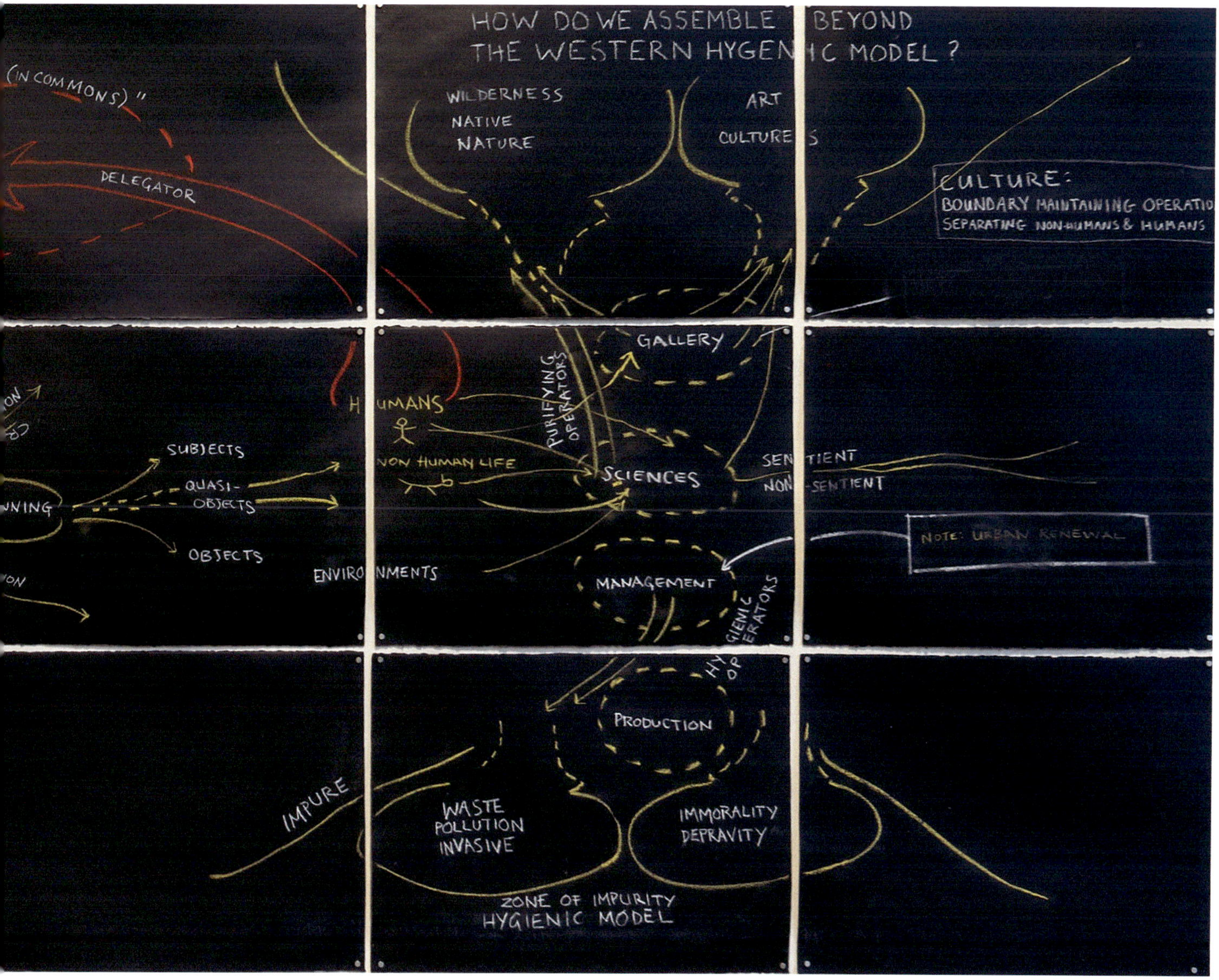
HOW DO WE ASSEMBLE BEYOND
THE WESTERN HYGENIC MODEL?
(IN COMMONS)"
DELEGATOR
WILDERNESS
NATIVE
NATURE
ART
CULTURES
CULTURE:
BOUNDARY MAINTAINING OPERATIO
SEPARATING NON-HUMANS & HUMANS
GALLERY
PURIFYING OPERATORS
HUMANS
SUBJECTS
QUASI-OBJECTS
OBJECTS
NON HUMAN LIFE
SCIENCES
SENTIENT
NON-SENTIENT
NOTE: URBAN RENEWAL
ENVIRONMENTS
MANAGEMENT
HYGIENIC OPERATORS
PRODUCTION
IMPURE
WASTE
POLLUTION
INVASIVE
IMMORALITY
DEPRAVITY
ZONE OF IMPURITY
HYGIENIC MODEL

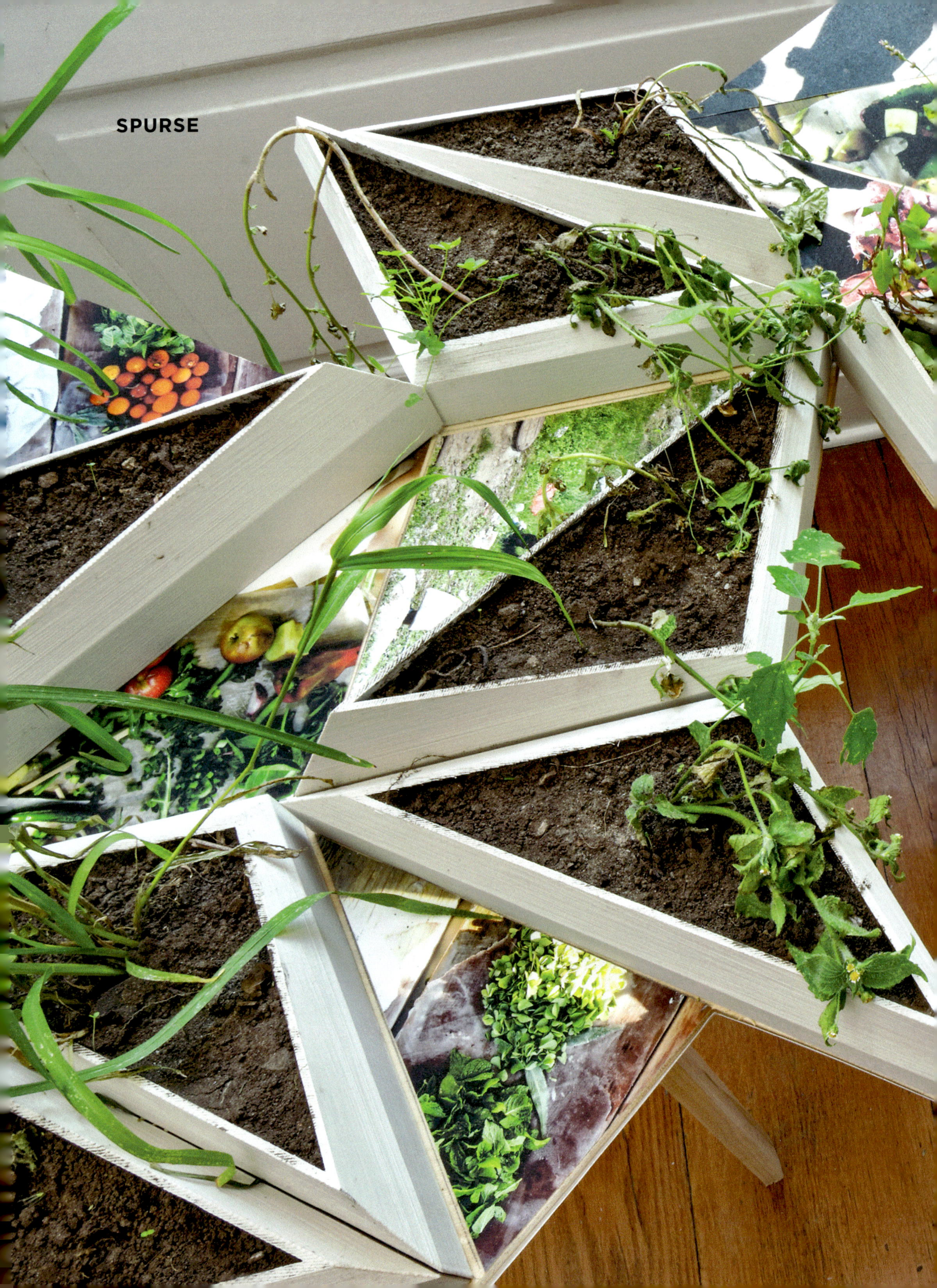

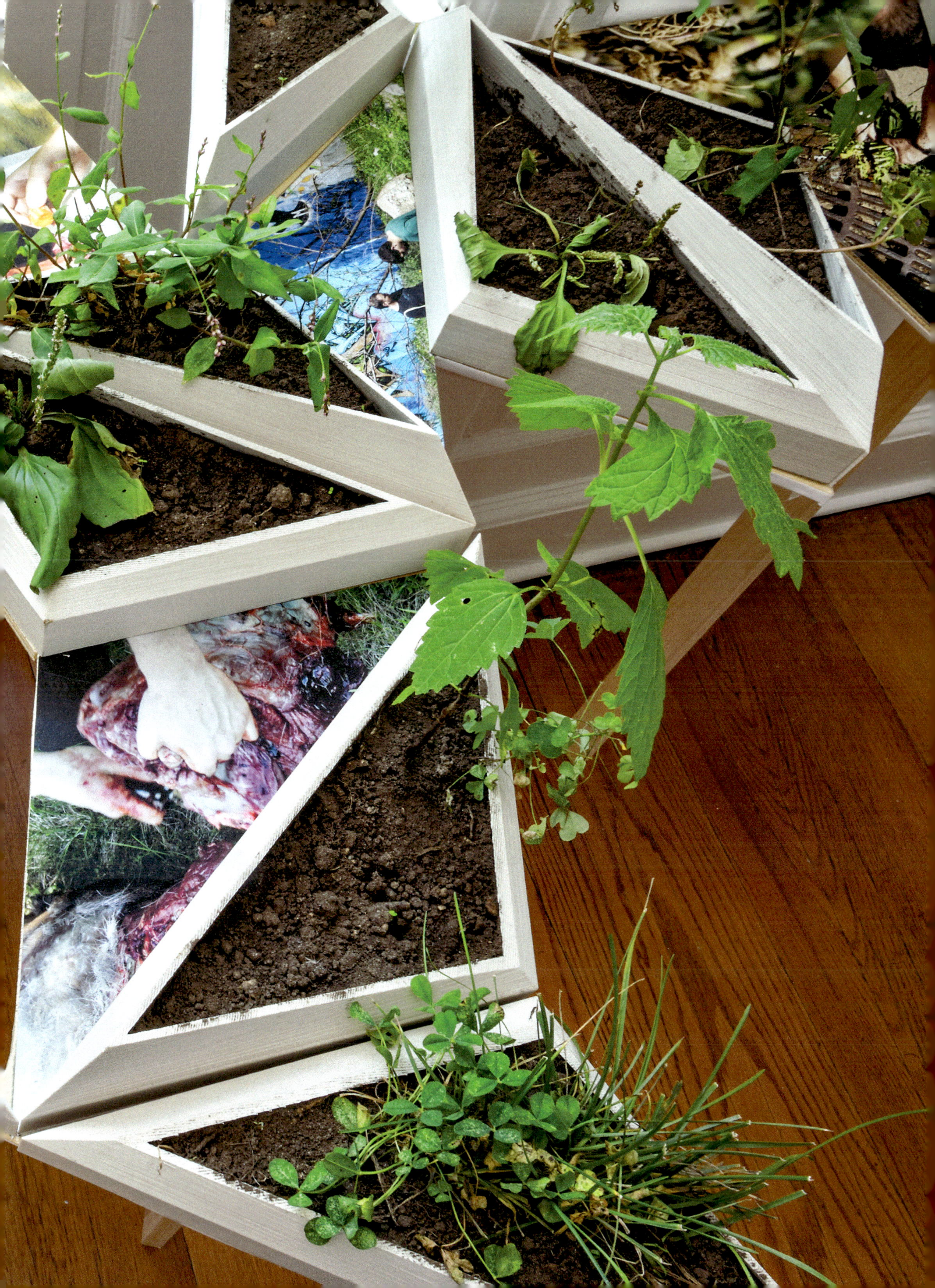

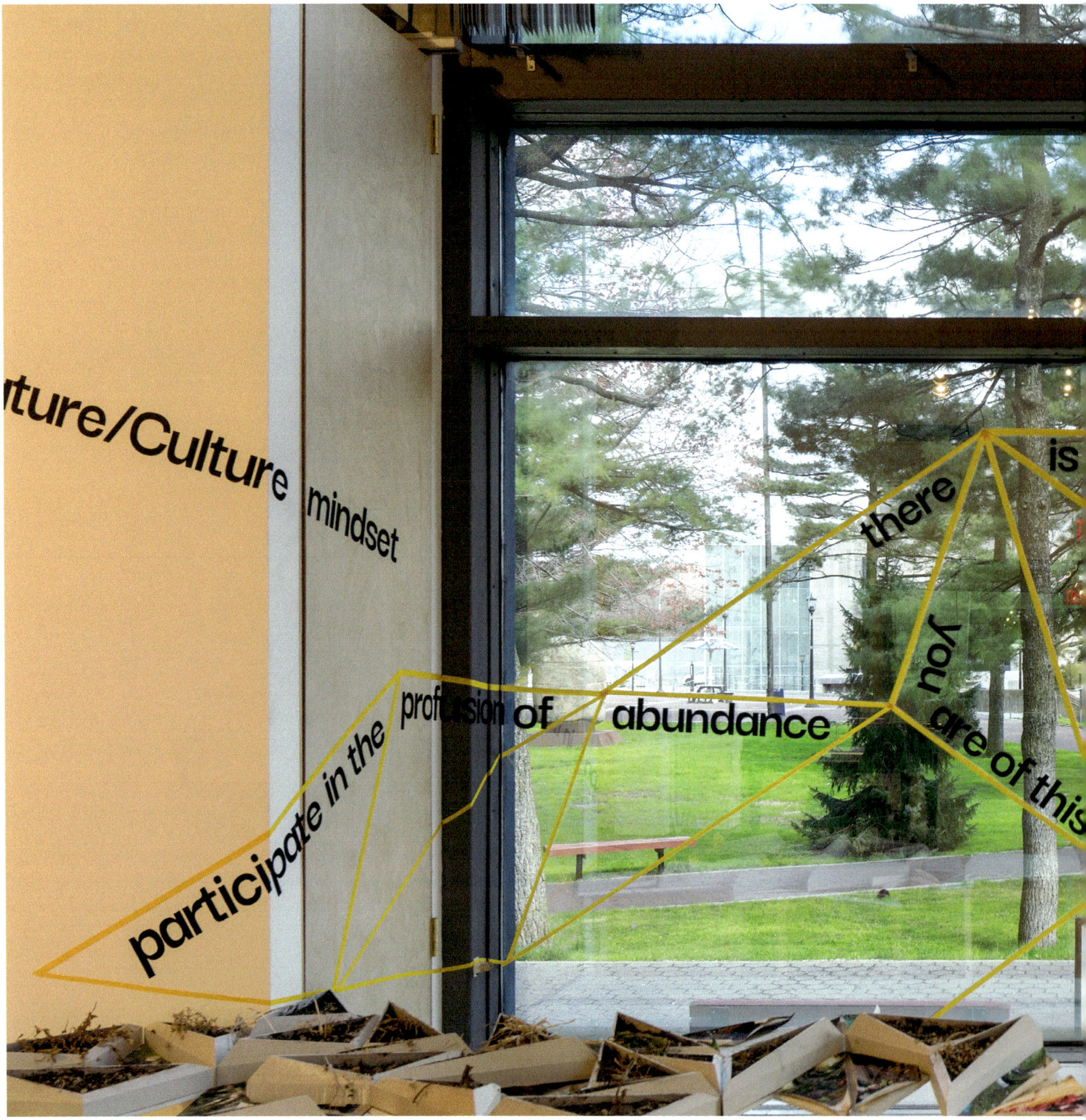
ture/Culture mindset
there
is
participate in the
profusion of
abundance
you are of this

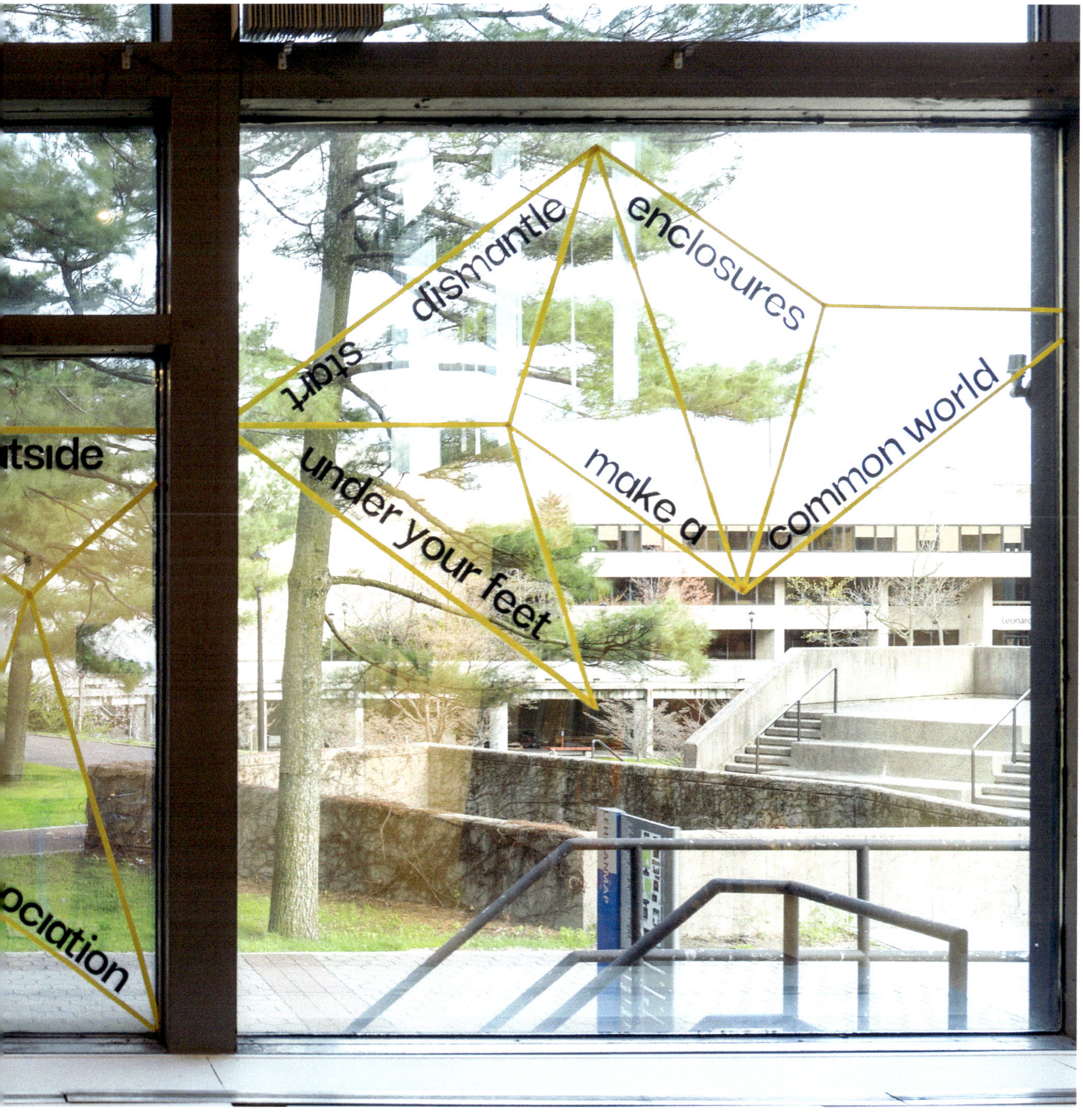
start
dismantle
enclosures
make a
common world
under your feet
tside
ociation

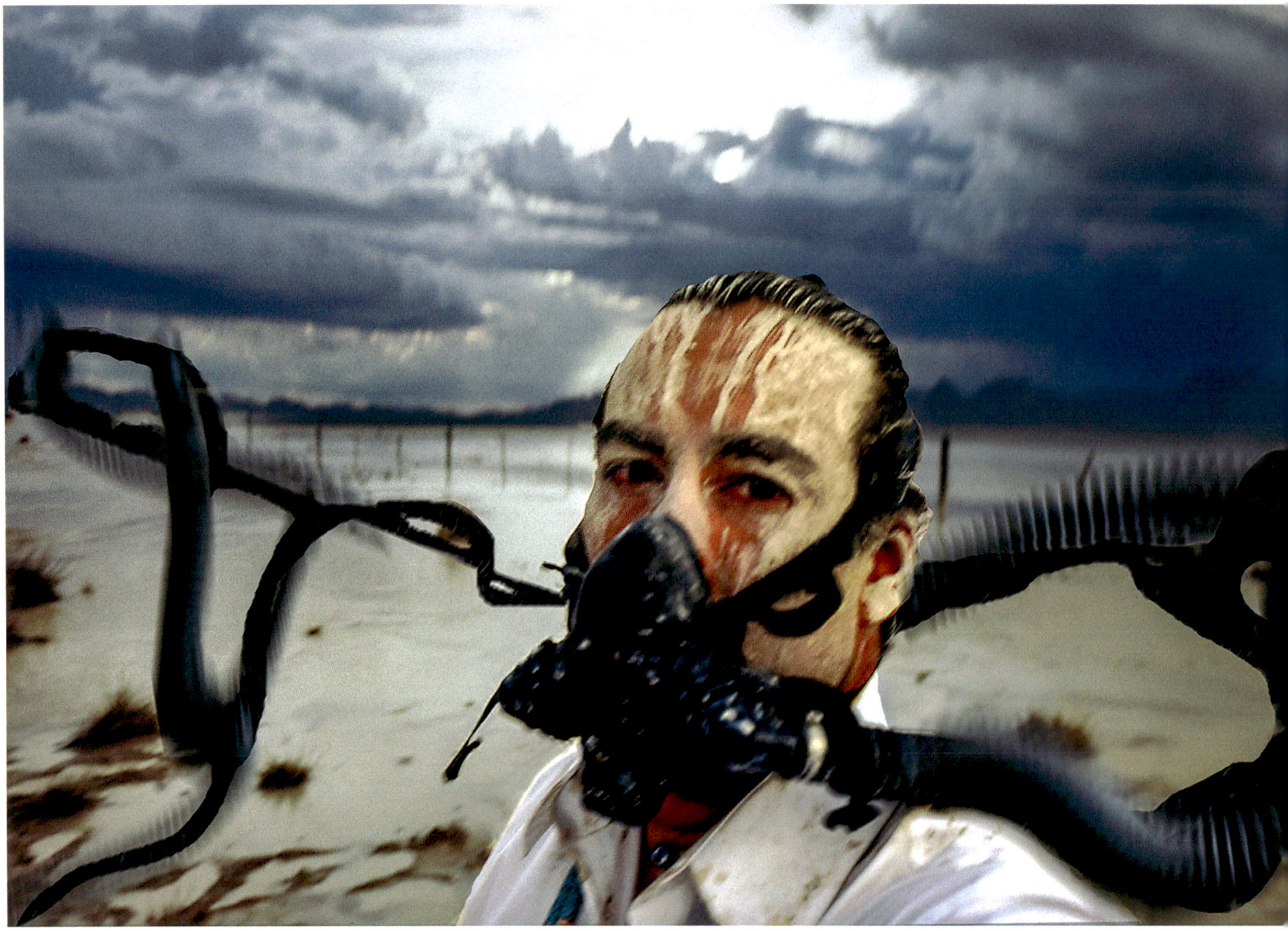

AUTO IMMUNE RESPONSE (AIR) NO. 5, 2004
Archive pigment print, 24 x 66 inches
Courtesy of the artist

Wilson is a Native American who spent his formative years growing up in the Navajo Nation. His ongoing series, *Auto Immune Response* (AIR), portrays the relationship between an imagined post-apocalyptic Diné (Navajo) man and the magnificent but toxic environment he inhabits. The *Auto Immune Response* is also this artist's reminder of the many autoimmune diseases, including lupus and rheumatoid arthritis that affect Native Americans. This environment in which the protagonist, his face covered with blood and dust, must wear a respirator, is so bleak that it might be the surface of the moon. Wilson describes his series as "an allegorical investigation of the extraordinarily rapid transformation of indigenous

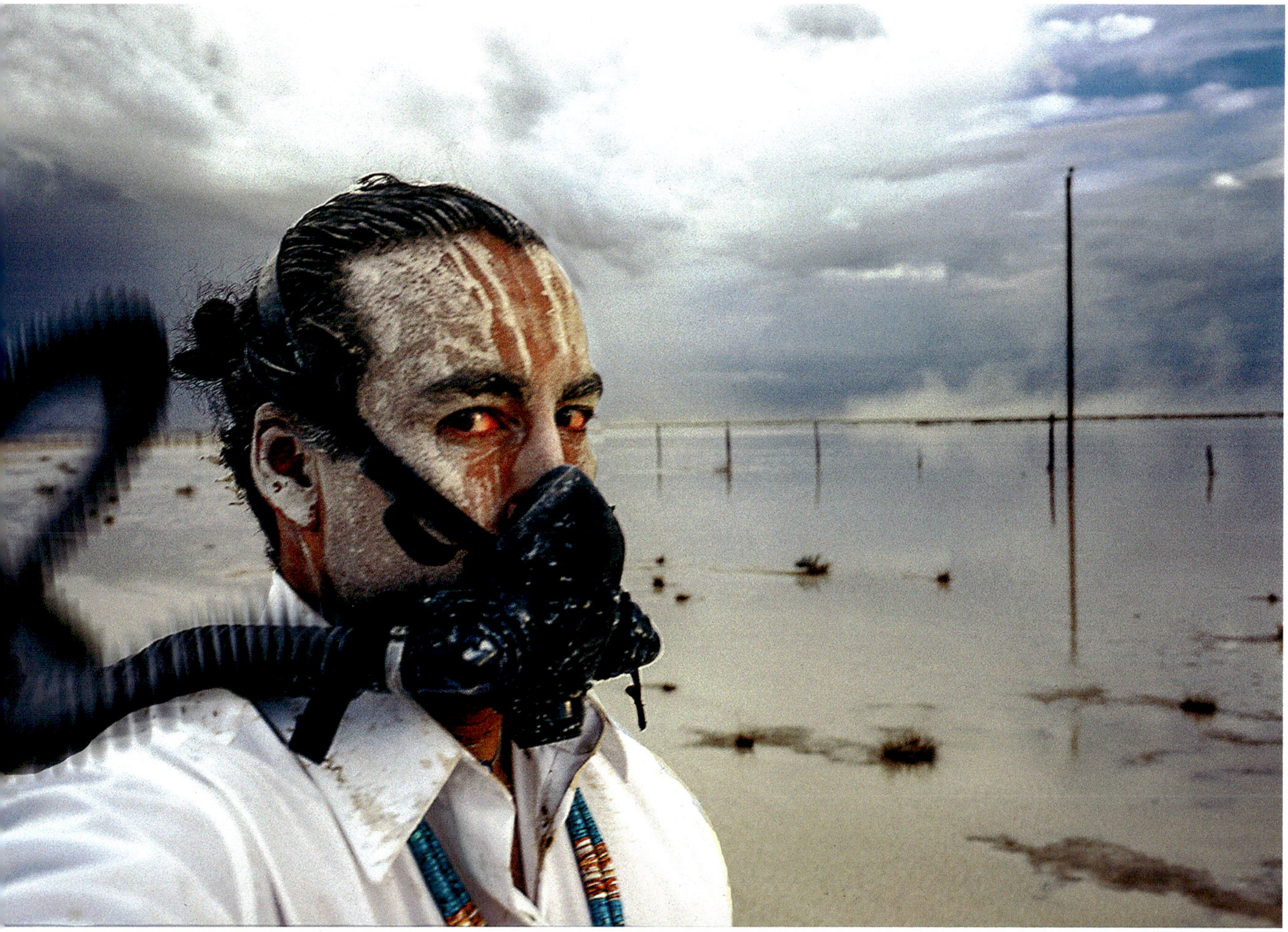

WILL WILSON

lifeways, the disease it has caused, and strategies of response that enable cultural survival." Wilson's powerful photographs tackle the interconnecting issues of Native American identity, contested land, colonization, health, and the melding of art and technology. He photographs himself in this landscape of the Navajo Nation, located across today's Arizona, New Mexico, and Utah. The land is the strong second character in Wilson's work, as the artist reminds us of its tortured history. Corporations and the U.S. government stripped its natural resources, particularly toxic uranium ore used to build the nation's atomic arsenal between World War II and the 1980s, Despite repeating his face to create a dramatic mirroring effect, Wilson, instead, creates an image of profound loneliness — a portrait of the artist in isolation.

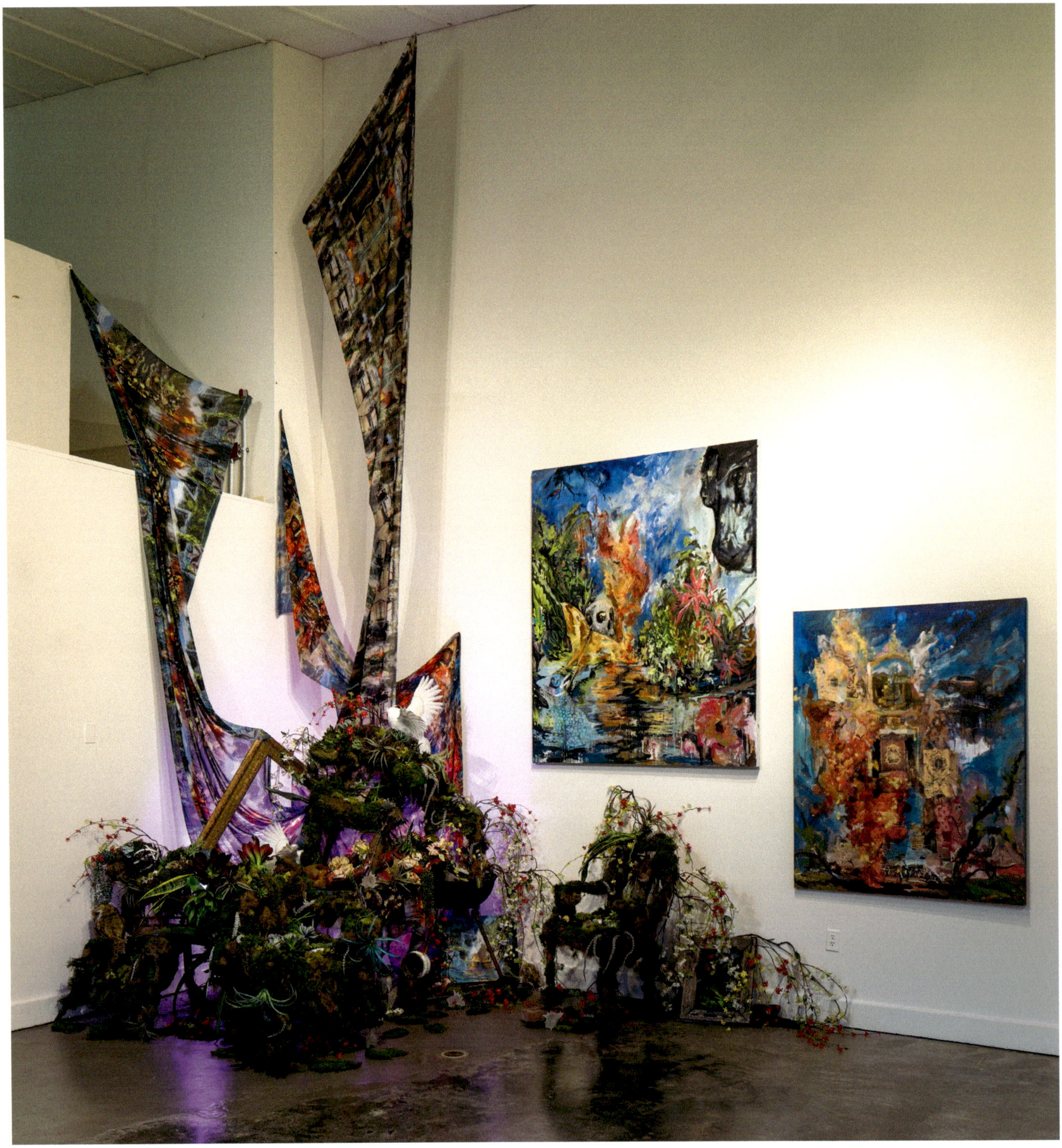

THE DINNER AFTER THE STORM, 2021
Chicken wire, repurposed furniture, plants, moss, and digitally printed chiffon
72 x 120 x 180 inches
Courtesy of the artist

NATALIE COLLETTE WOOD

Wood investigates the growing dichotomy between the natural world and the pressures of urban development in an age of climate change. Her latest installation, *The Dinner After the Storm*, suggests a bucolic backyard barbeque gone terribly awry and recalls the freakish weather events that global warming brings, such as the "remnants" of Hurricane Ida during the fall of 2021, which killed nearly 50 people in the New York City area in a flash flood. Wood grew up in a modern city, Las Vegas, a pinnacle of discordance between the natural and manmade worlds. The desert that surrounds this city deeply impacted Wood. Both beautiful and ferocious, it shaped her respect for nature. Wood tempers her awareness of the chaos nature can cause by her interest in its beauty. Deeply inspired by historical art, she cites the storm scenes of 19th-century British landscape painter J.M.W. Turner as an important influence in her works on canvas. One of the best-known aspects of her installations is her use of chairs, often baroque design that she covers with succulents, moss, and weeds suggesting the poetry of decay. She often presents the chairs askew, signaling a toppled grandeur or false nostalgia. Wood is a 2008 graduate of the M.F.A. program at Lehman College.

NATALIE COLLETTE WOOD

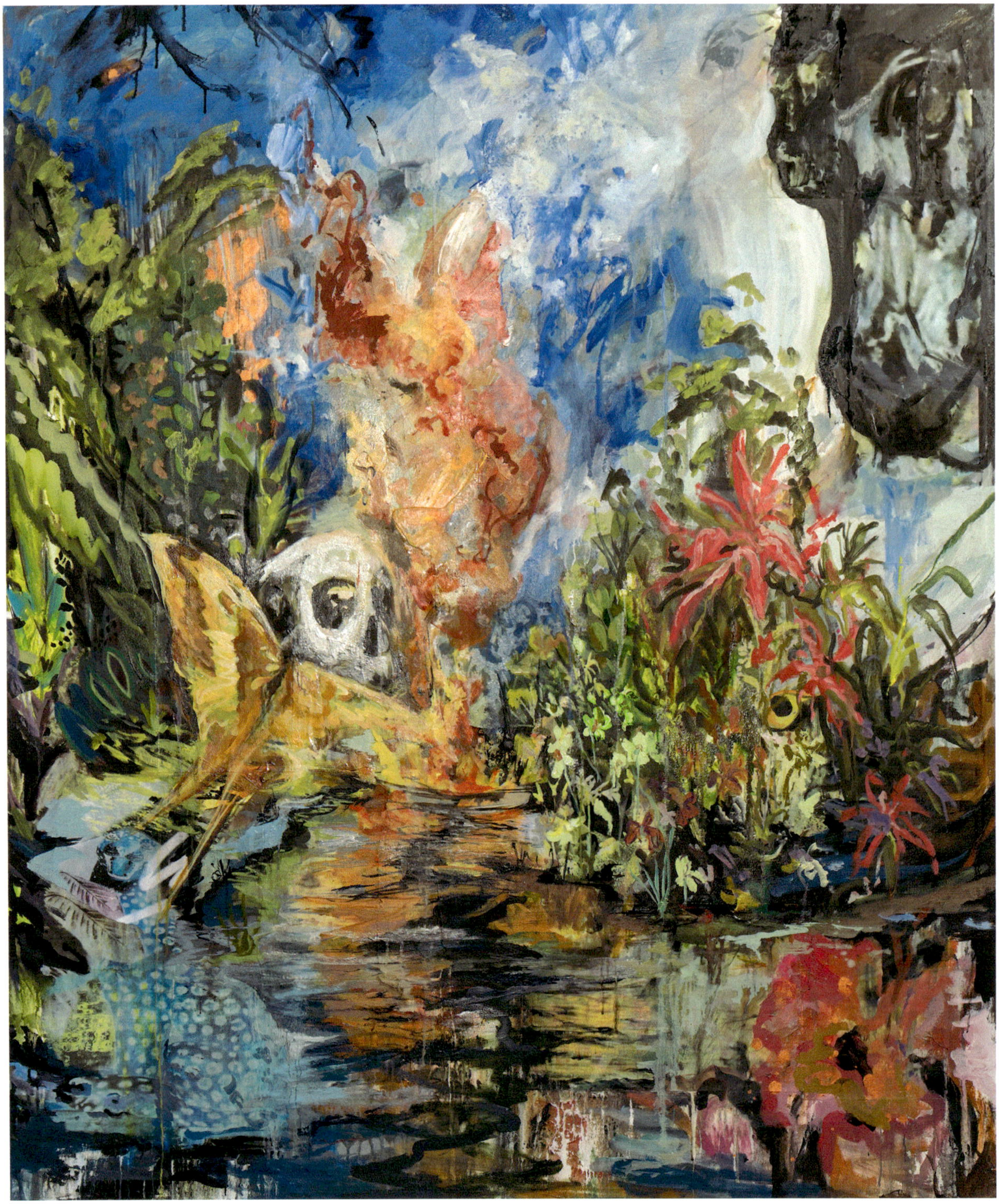

RIVER META, 2021
Acrylic oil and printed paper pasted on canvas
60 x 72 inches
Courtesy of the artist

BACKYARD BBQ, 2018
Acrylic oil and printed paper pasted on canvas
48 x 60 inches
Courtesy of the artist

RIGHT

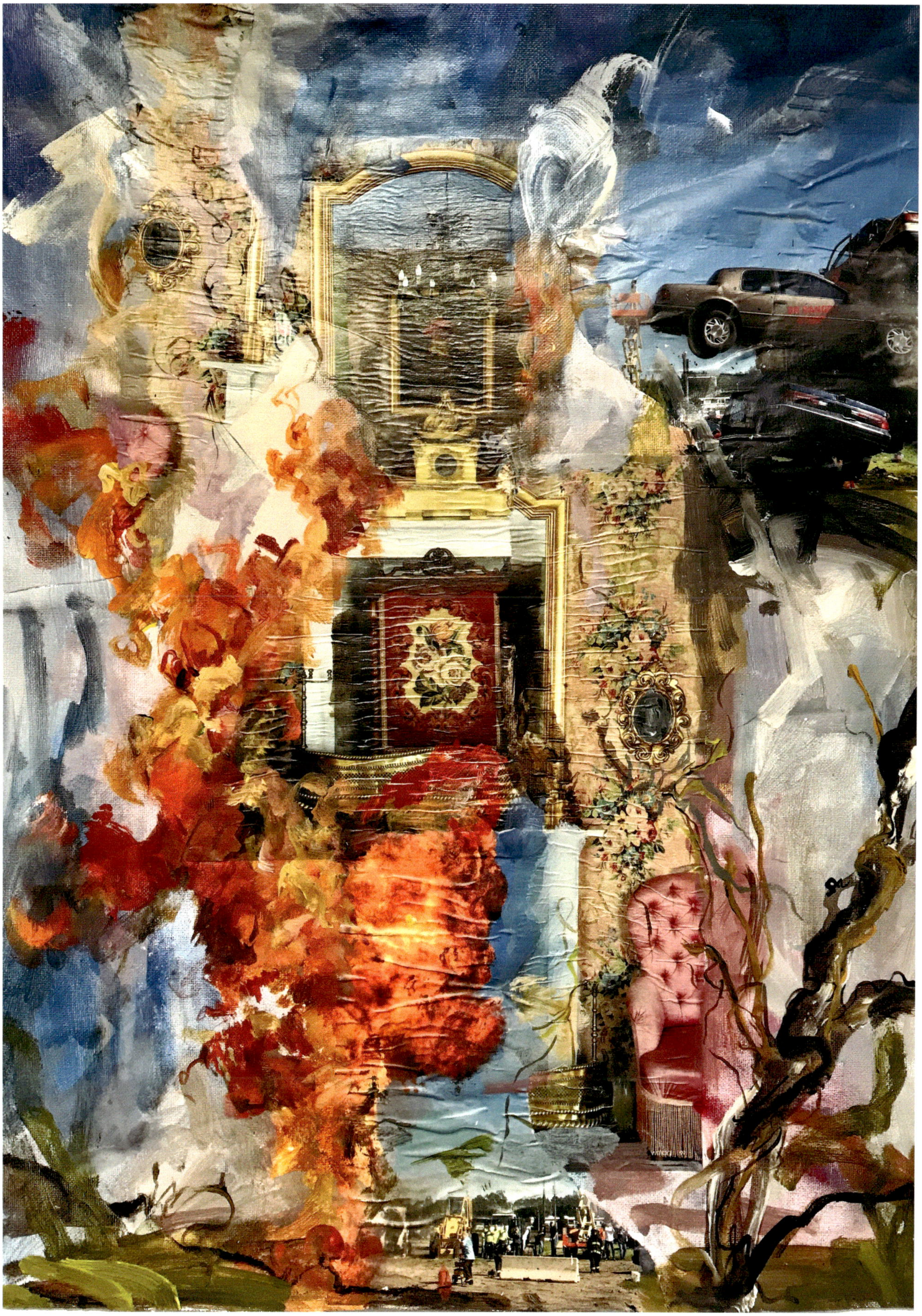

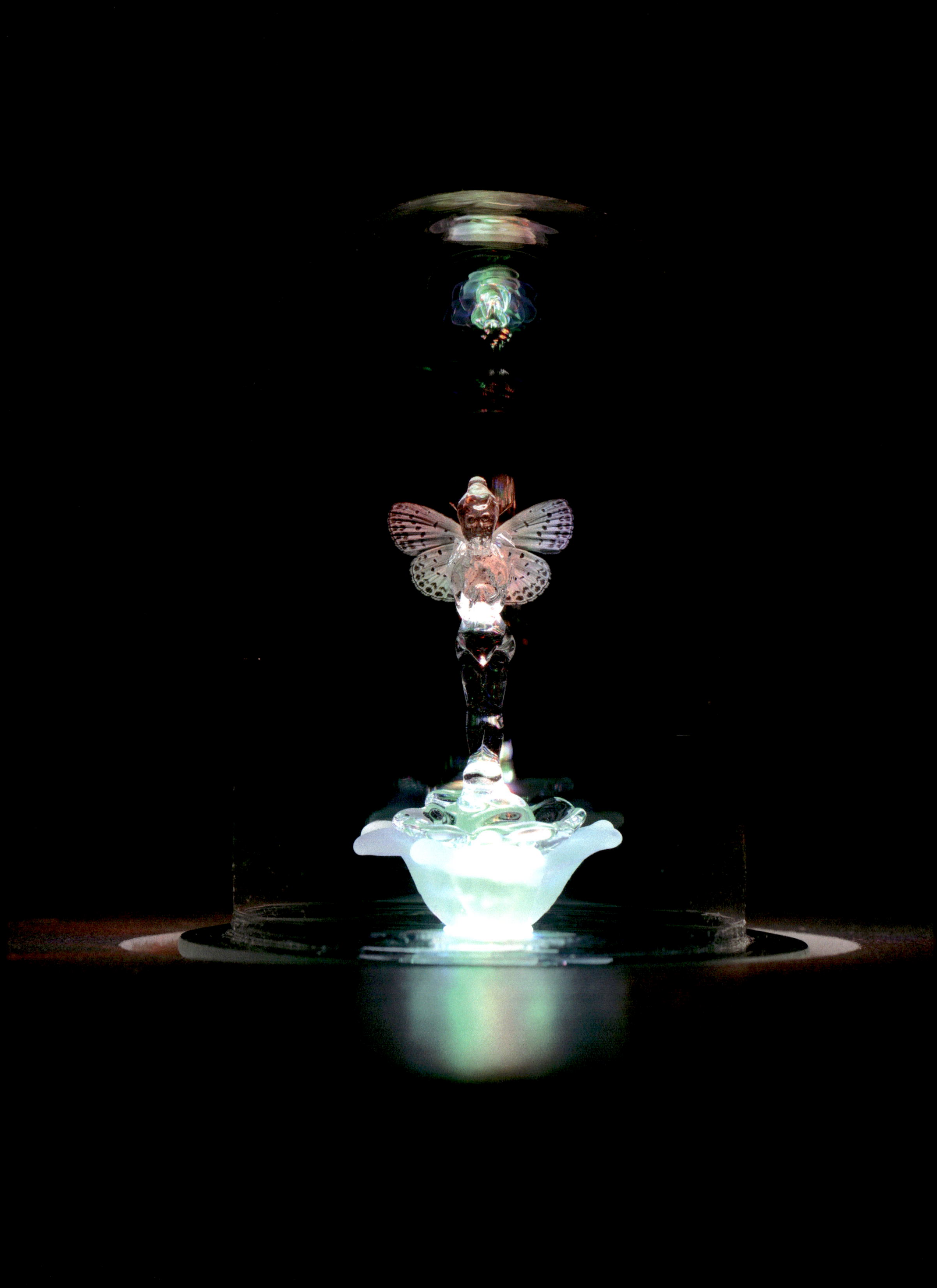

THREE WISHES, 2014
Three sculptures, each 6 Å~ 3 inches
Glass figurines, butterfly specimens, glass domes, music boxes
Courtesy of the Artists and Mizuma Art Gallery

KEN + JULIA YONETANI

Ken + Julia Yonetani's work delves into nature and the spiritual today, engaging the five senses and the theatricality of performance. *Three Wishes*, a seemingly whimsical presentation of illuminated butterflies, packs a punch and draws inspiration from Walt Disney's television series *Wonderful World of Color* and its 1957 episode entitled, "Our Friend the Atom." In this show a genie offers three wishes: limitless energy, magic to conjure food, and health and peace, all suggesting the benefits of atomic power. But the artists of *Three Wishes*, in a collaboration with scientists investigating the effects of the 2011 nuclear disaster at Japan's Fukushima Daiichi power plant, created three sculptures of butterflies 57 years after Disney's enthusiastic introduction of the atom. Their sculptures show butterflies, unwitting victims of nuclear energy gone wrong. Evident in them are mutations and high death rates from radiation poisoning. The rotating insects create a distressing viewpoint from which to ponder Disney's iconic song, "It's a Small World," a sober reminder of our global toxic connectivity.

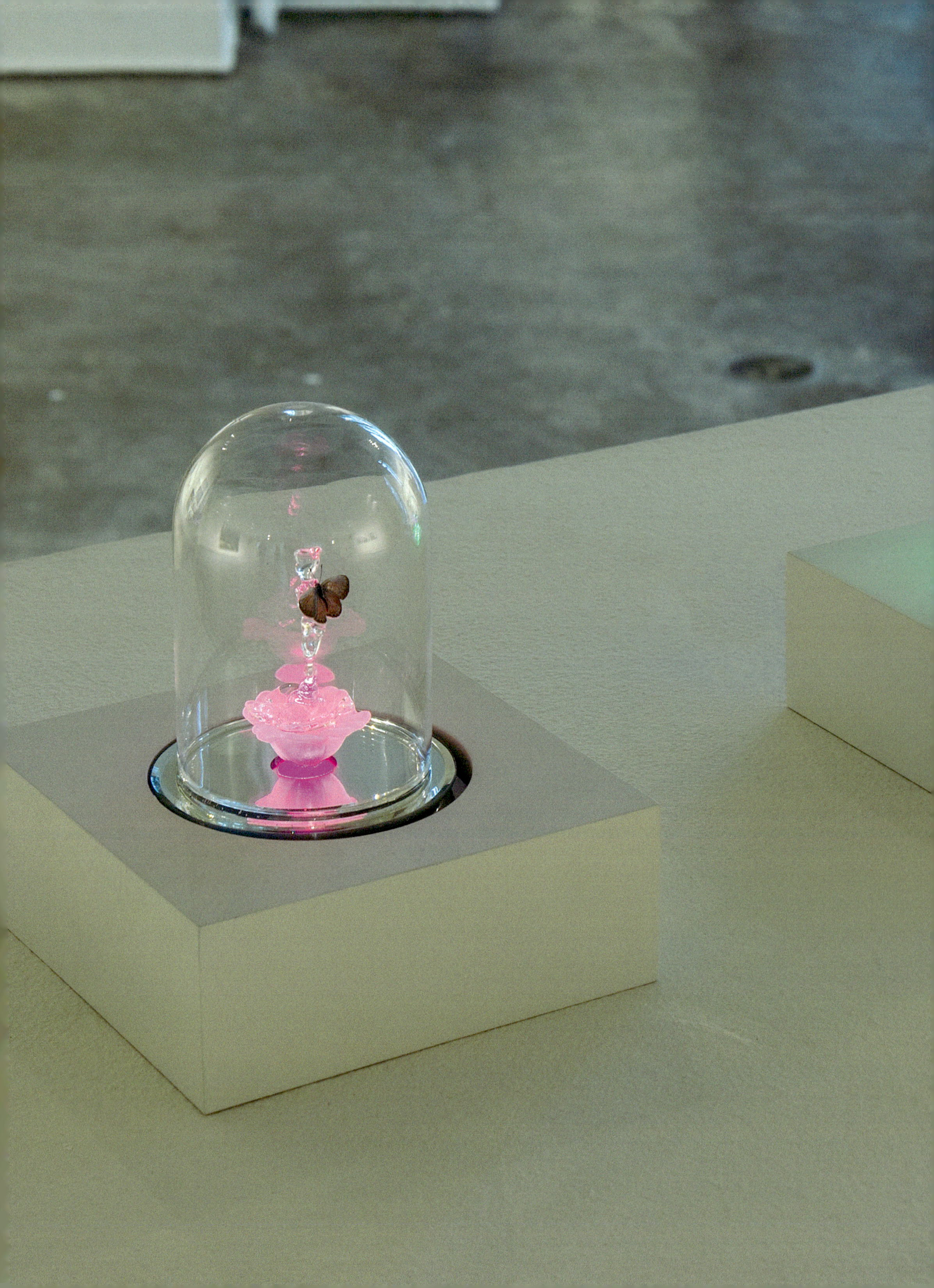

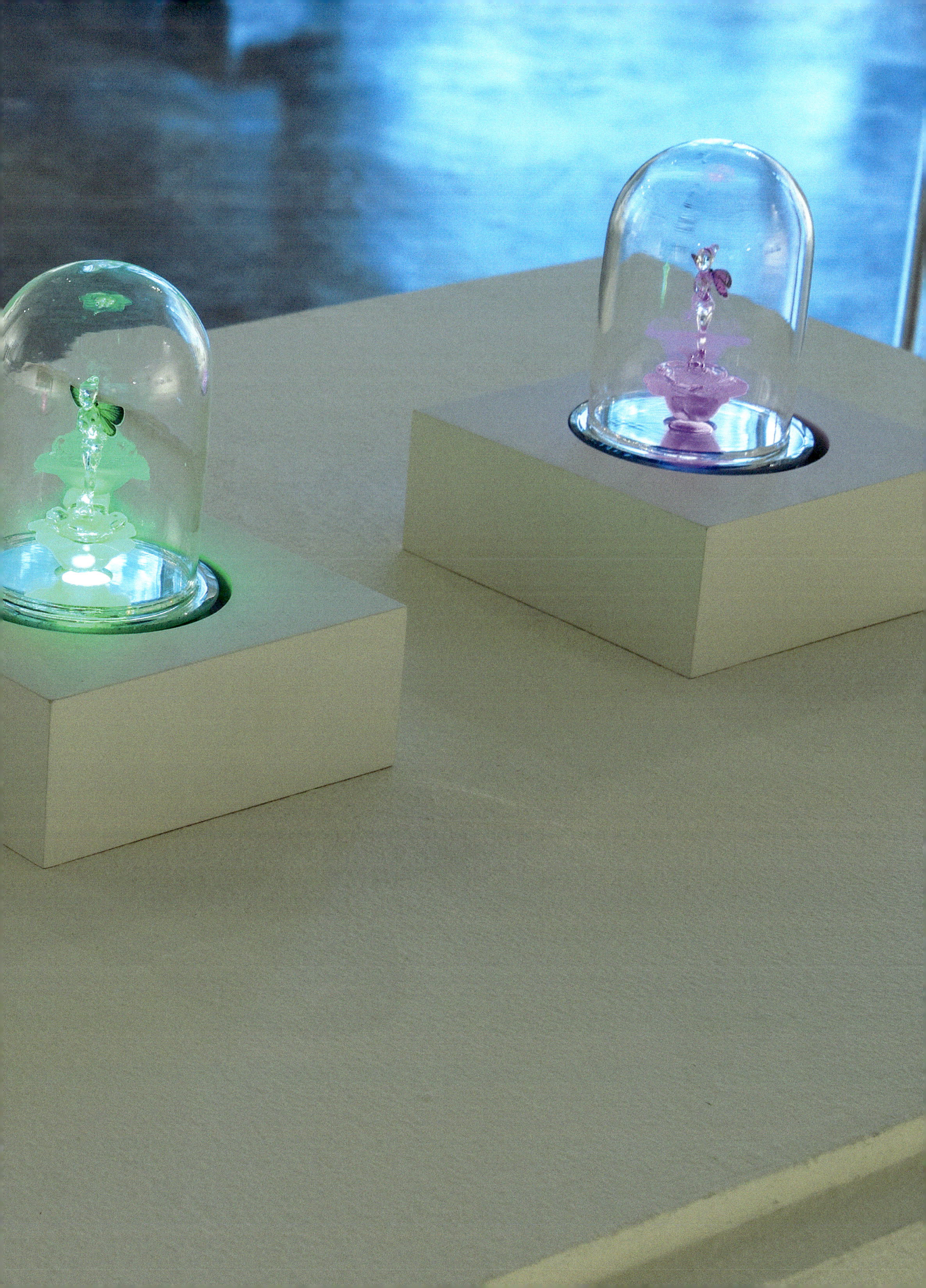

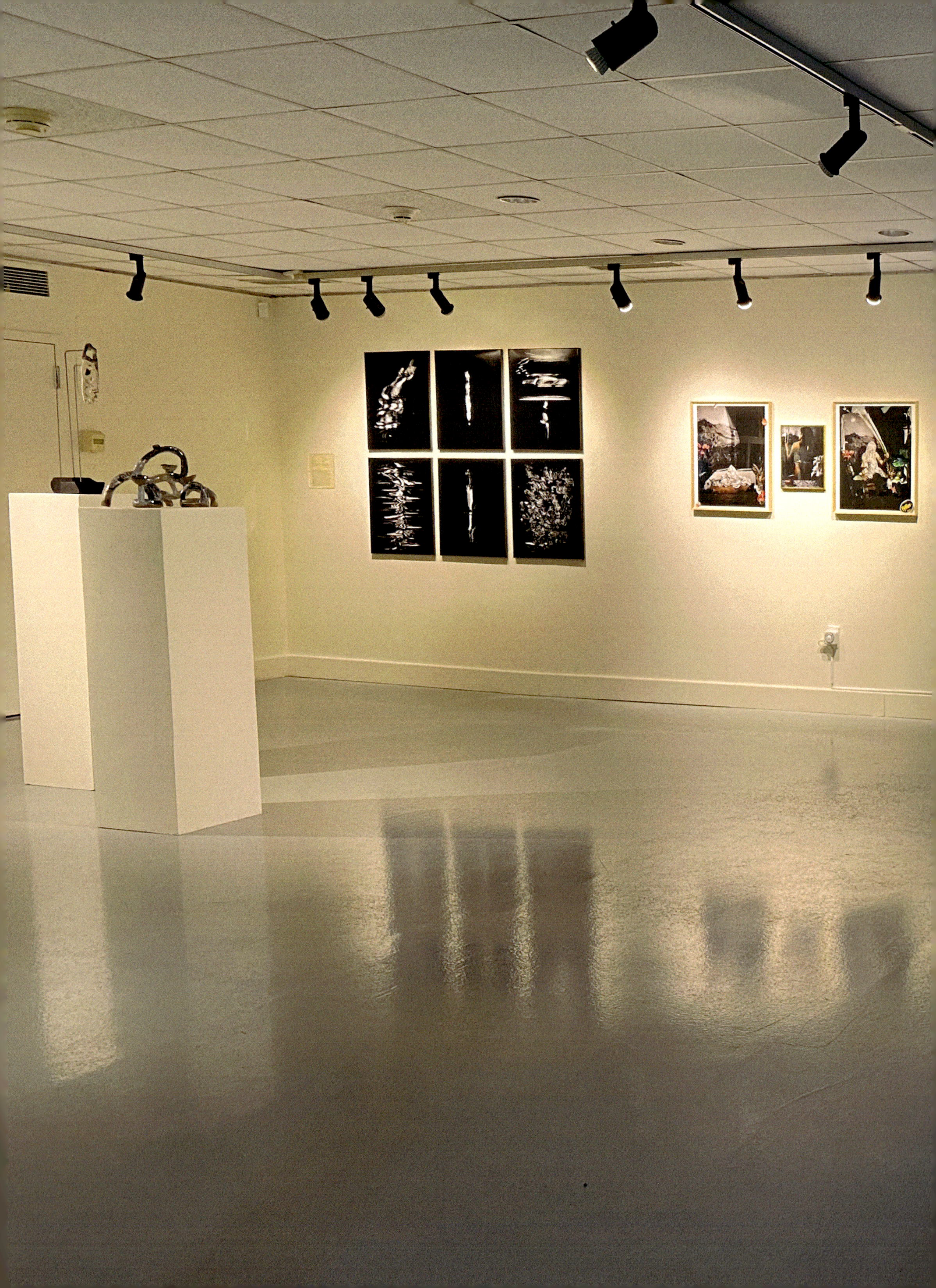

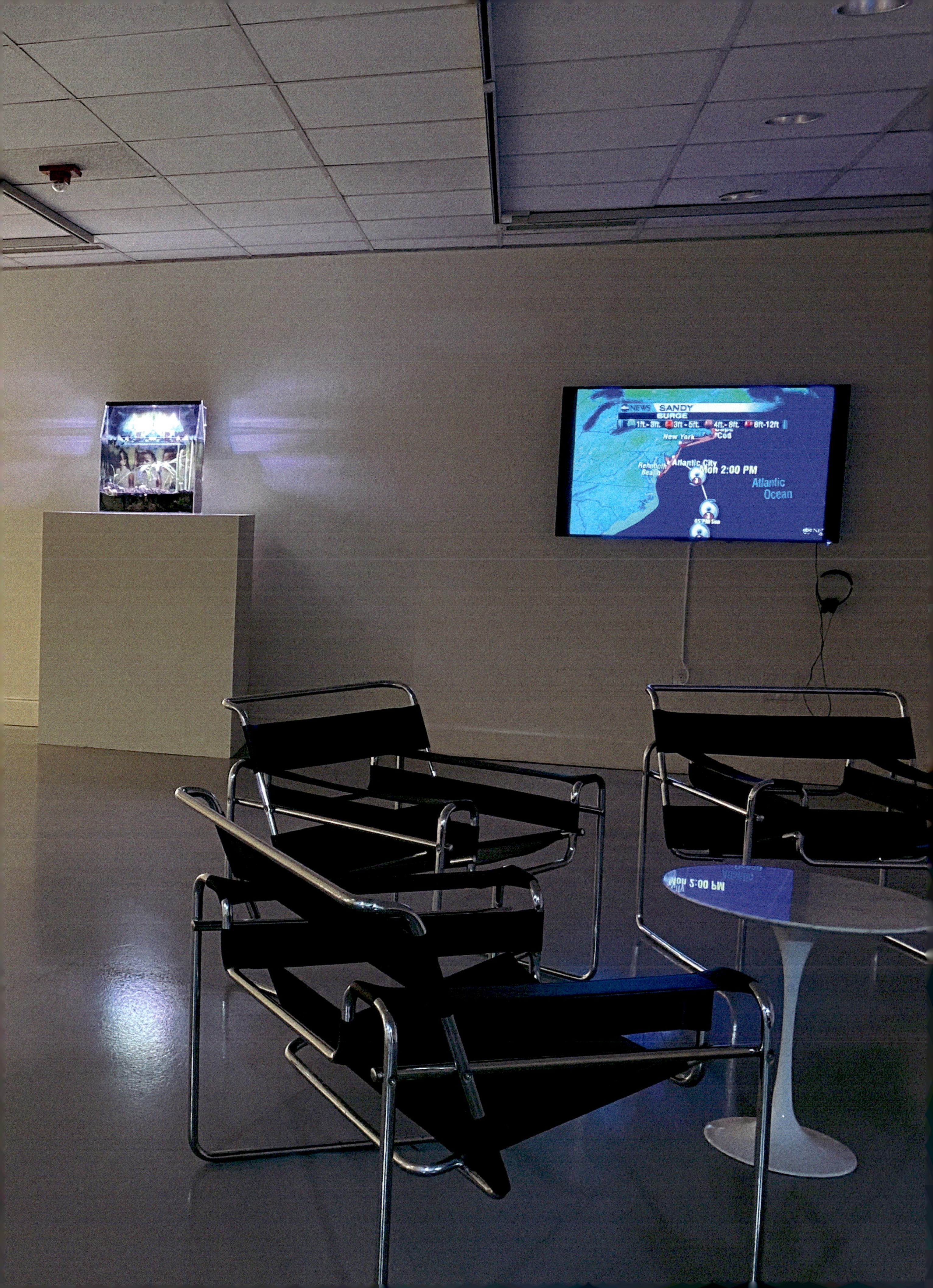
SANDY
SURGE
New York
Atlantic City
Mon 2:00 PM
Atlantic Ocean

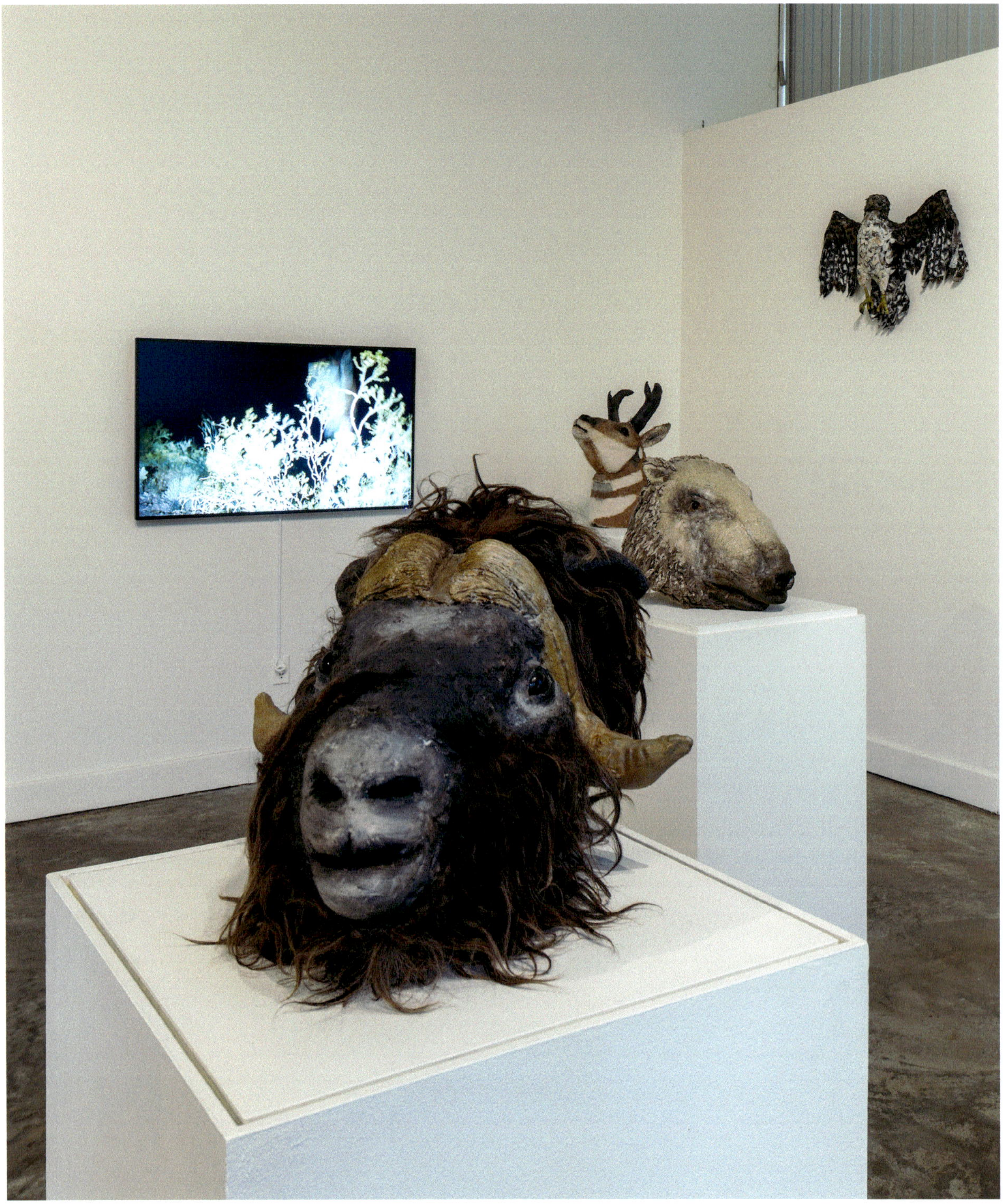

LEHMAN COLLEGE ART GALLERY

Always free to the public, Lehman College Art Gallery has been serving the interests of our diverse audience from the Bronx and New York City since 1984. The gallery specializes in thematic group exhibitions that bring together famous artists with emerging talents. Education is an integral component of the Gallery's programming and provides the basis of community outreach to more than 25,000 individuals a year — from young students to senior citizens.

Staff

Bartholomew F. Bland, Executive Director
Jessica Fernandez De Jesus, Gallery Assistant
Mary Ann Siano, Grants Associate
Deborah Yasinsky, Curator of Education

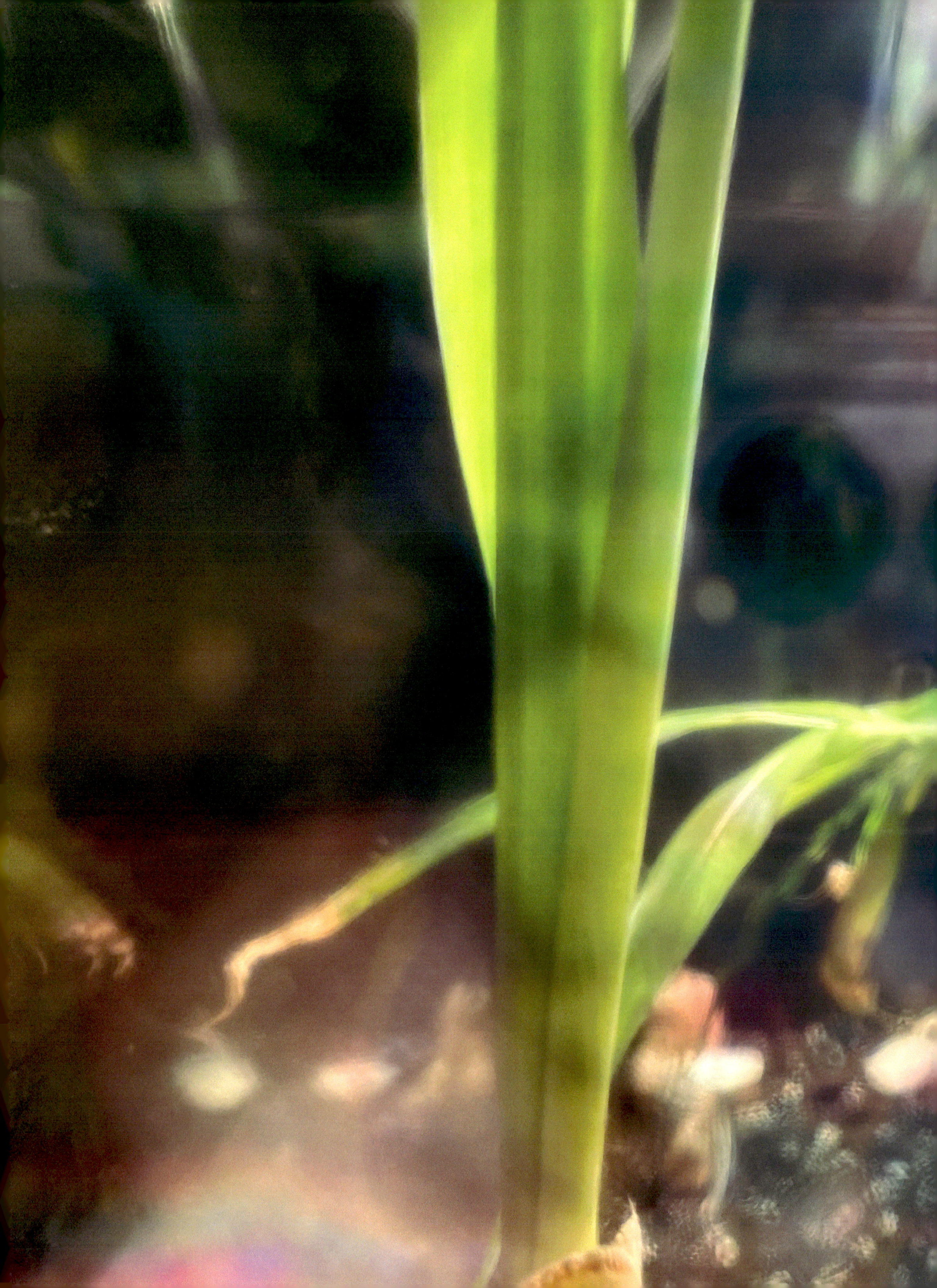

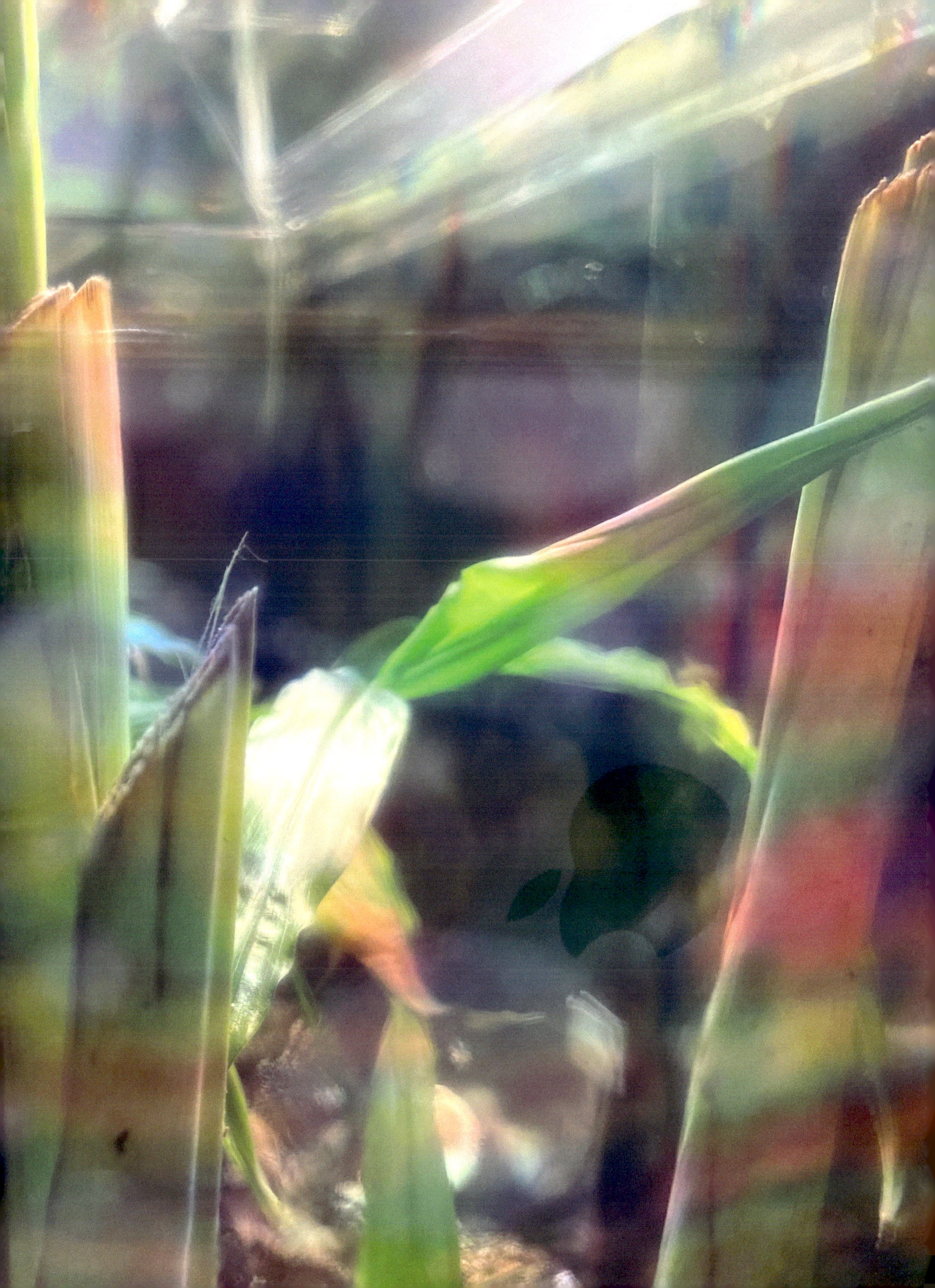

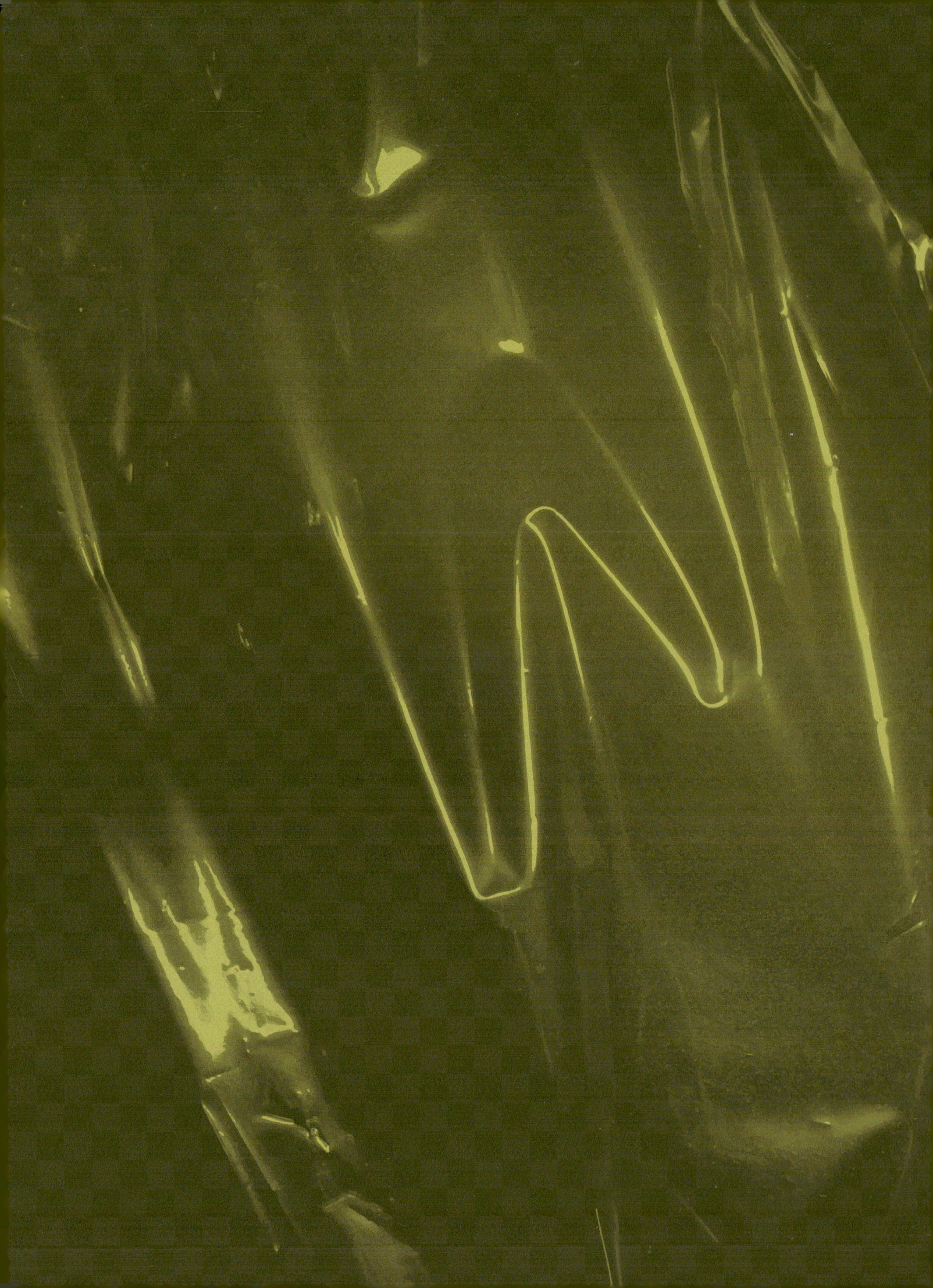